Palgrave Studies in Educational Media

There is no education without some form of media. Much contemporary writing on media and education examines best practices or individual learning processes, is fired by techno-optimism or techno-pessimism about young people's use of technology, or focuses exclusively on digital media. Relatively few studies attend – empirically or conceptually – to the embeddedness of educational media in contemporary cultural, social and political processes. The **Palgrave Studies in Educational Media** series aims to explore textbooks and other educational media as sites of cultural contestation and socio-political forces. Drawing on local and global perspectives, and attending to the digital, non-digital and post-digital, the series explores how these media are entangled with broader continuities and changes in today's society, with how media and media practices play a role in shaping identifications, subjectivations, inclusions and exclusions, economies and global political projects. Including single authored and edited volumes, it offers a dedicated space which brings together research from across the academic disciplines. The series provides a valuable and accessible resource for researchers, students, teachers, teacher trainers, textbook authors and educational media designers interested in critical and contextualising approaches to the media used in education.

More information about this series at
http://www.palgrave.com/gp/series/15151

Patrick Bettinger
Editor

Educational Perspectives on Mediality and Subjectivation

Discourse, Power and Analysis

palgrave
macmillan

Editor
Patrick Bettinger
Center for Education and Digital Transformation
Zurich University of Teacher Education
Zurich, Switzerland

ISSN 2662-7361 ISSN 2662-737X (electronic)
Palgrave Studies in Educational Media
ISBN 978-3-030-84342-7 ISBN 978-3-030-84343-4 (eBook)
https://doi.org/10.1007/978-3-030-84343-4

This Palgrave Pivot imprint is published by the registered company Springer Nature Switzerland AG.
The registered company address is: Gewerbestrasse 11, 6330 Cham, Switzerland

Foreword

There is no education without some form of media. The field of educational media is a growing area of interest, as policy papers on the 'digital agenda', the rapid expansion of conferences and events on educational technology, and the range of recent books and articles on education and media show. Educational media are crucial to producing knowledge and shaping educational practices. Much interest in recent years has, however, focused on which digital technology should be used, on the benefits and risks of using 'new' technologies, on good practices for teaching and learning with media, or on how media supports individual learning processes.

While these are important issues, there has been too little attention to the socio-political contexts in which educational media are developed and used, to the societal discourse in these media, and the relations enacted when we use educational media. Conflicts over the contents of textbooks and curricula, flaring up regularly in the daily news across the world, illustrate how many stakeholders are invested in sharing their particular understandings of our (shared) past, the current society and potential imagined futures with the younger generation. Policymakers, politicians and activists regard educational media as important tools which not only foster young people's media skills and world knowledge, but also shape which ways of living are considered desirable or even legible.

Today, an increasing number of studies see educational media as a highly contested and thus crucially important cultural site. This research considers media in their contexts and takes a carefully critical or generative approach to societal concerns. The *Palgrave Studies in Educational Media* series brings together current research on educational media by focusing on three issues:

First, it foregrounds studies which attend—empirically or conceptually—to how deeply embedded textbooks and other educational media are in contemporary cultural, social and political processes, and to the historicity of the media used in education.

Second, it brings together vibrant and dynamic scholarship drawing on a range of disciplines—including sociology, history, cultural studies, computer science, memory studies, information science, media studies, education and cognitive science—to build and develop insights generated and exchanged across disciplinary boundaries.

Third, although the field of educational media studies has turned its attention to digital technologies, taking a closer look at today's educational practices, it is clear that (1) they are by no means predominantly digital, and simultaneously (2) 'postdigital' practices abound in which the digital is no longer seen as new or innovative, but is integrated with other materials in daily teaching and learning. Empirical observations of education around the globe demonstrate the reach and visibility of a broad range of media (textbooks, blackboards, LEGO™, etc.), as well as the postdigital blending of digital and non-digital media in contemporary educational settings.

Palgrave Studies in Educational Media aims to address these three issues in an integrated manner. The series offers a dedicated space which encourages dialogue across disciplines. It showcases both empirical and theoretical work on educational media which understands these media as a site of cultural contestation and socio-political force. The focus lies primarily on schools. The series is interested in both local and global perspectives, in order to explore how educational media are entangled with broader debates about continuity and change in today's society, about classroom practices, inclusions and exclusions, identifications, subjectivations, economies and global political projects.

The present volume, edited by Patrick Bettinger, is a long overdue collection that fills a research lacuna around the intersection of research on subjectification and mediality from an education sciences perspective. While research in these areas to date (usually in either one or the other) has tended to follow a more binary (subject vs. object) or anthropocentric (human vs. technology) approach, the chapters of this book conceive of subjectification and mediality as contingent and hybrid processes that instigate, are products of, and constitute not only each other but also the socio-technical aspects of school-related digital media, including both human and non-human factors. As Bettinger explains in his introduction, this innovative perspective is of key importance to the education sciences. It is often at the cross-section of language, discourse and subjectification via digital media in an educational setting that power relations can shift or settle. Hitherto centered around a competency- and output-orientated learning paradigm that 'uses' media as learning 'instruments' in order to fulfill pre-defined objectives, power dynamics in the classroom are seen from a fresh perspective in the performative interrelationship of subjectification and mediality. At this intersection, concepts of power and knowledge can be reconsidered in the educational context with all its hybrid, multi-layered complexities. We are delighted to present this innovative and cutting-edge contribution to educational media research in this, the sixth book of our series.

Brunswick, Germany
January 2021

Eckhardt Fuchs

subjectification and meaning as communicative and [illegible] processes. [illegible] instance, are products of and constituting not only [illegible] socio-technical aspects of school-related digital [illegible] of the human and non-human factors. As it emerges [illegible] contribution on this innovative perspective of assigning importance to [illegible] It is often at the intersection of human and [illegible] via digital media in any educational setting that [illegible] settle. This [illegible] centered around accompanying [illegible] learning paradigm that uses media as learning [illegible] fulfil pre-defined objectives, power dynamics in [illegible] from a fresh perspective on the performative interrelation [illegible] cation and mediation. At this intersection, concepts of [illegible] edge can be reconsidered in the educational context [illegible] multi-layered complexities. We are delighted to [illegible] and curious inter-disciplinary contributions to educational media [illegible] sixth book of our series.

Brunswick, Germany [illegible]
January 20[illegible]

Contents

Notes on Contributors

Patrick Bettinger is Professor for Media Education at the Zurich University of Teacher Education, Switzerland. His research focusses on qualitative methods and methodologies in educational media research, educational media theory and the design of learning processes with digital media.

Valentin Dander is an education researcher and Professor of Media Education at the University of Applied Sciences Clara Hoffbauer in Potsdam, Germany. His research interests include media and civic education, philosophy and sociology of (media) education, digitisation and datafication in society, media critique and open education.

Viktoria Flasche is research assistant at the Department of Education with a focus on culture and aesthetic education at the Friedrich-Alexander-University-Erlangen-Nuremberg. Her research interests are in the field of youth research with a focus on processes of subjectivation—especially with regard to social media—and reconstructive approaches in social and art studies.

Britta Hoffarth Dr. phil., Professor of Gender and Educational Cultures and head of the Centre for Gender Studies, University of Hildesheim. Recent works deal with affectivity and politics, intersectional perspectives on digitality or television series and praxeological approaches to the body. Her main areas of work are: Gender, youth, migration, body, media, materiality.

Felicitas Macgilchrist heads the 'Media|Transformation' department at the Georg Eckert Institute for International Textbook Research, Braunschweig, and is Professor of Media Research at the University of Göttingen's Institute of Educational Science, Germany. She researches how people try to change education through technology, and how their design decisions have intended and unintended consequences.

Sabrina Schenk is a Postdoc Researcher with a focus on Educational Philosophy at the University of Frankfurt, Germany. Her research interests include contemporary forms of subjectivity and subjectivation, digitisation and datafication in society, politics and protests and their relevance for political education.

List of Figures

CHAPTER 1

Educational Perspectives on Mediality and Subjectivation: Introduction

Patrick Bettinger

Abstract The concept of the subject has long been a central construct of the social sciences, cultural studies and the humanities. While the philosophical roots of the concept go back to antiquity, new discourses have developed in recent years that critically question and further develop concepts such as subject or subjectivation. In addition to theoretical strands of discussion, the focus is increasingly on the empirical possibilities of subjectification research. It is becoming apparent that the constitutive power of digital mediality—also from the perspective of educational science—is playing an increasingly significant role in these contexts. The introductory chapter presents a brief outline of these developments and provides a first insight into the contributions in this volume.

Keywords Research • Power • Mediality • Digital culture • Relationality

P. Bettinger (✉)
Center for Education and Digital Transformation, Zurich University of Teacher Education, Zurich, Switzerland
e-mail: patrick.bettinger@phzh.ch

P. Bettinger (ed.), *Educational Perspectives on Mediality and Subjectivation*, Palgrave Studies in Educational Media,
https://doi.org/10.1007/978-3-030-84343-4_1

Investigation of 'the subject' is an interdisciplinary endeavour which for many years now has constituted an influential component of the fundamental discourse underlying the social and cultural sciences. Rooted in seventeenth- and eighteenth-century philosophy, our attempts to define 'the subject' can equally stand as proposed answers to the question of how we arrive at the knowledge we have. The discourses that have unfolded in this context have taken two distinct lines, with rationalism guiding the European debates and empiricism at the heart of the Anglo-American approach, each scaffolded by specific variants of theories of knowledge. Both strands have presented concepts of the subject in the process of acquiring knowledge and of its status in that process. Examining these ideas, we note divergent views on, *inter alia*, the extent of the subject's involvement in the act of knowing, with rationalism positing an active and empiricism a more passive role (Beer, 2014, p. 215). The term 'subject' itself is ambiguous in connotation, referencing both a singled-out status and a condition of subjugation (Reckwitz, 2006, p. 9). The classical conception advanced in continental Europe defines the subject as "a self-determined, self-transparent entity of knowledge and of—moral, interest-led or creative—action" (Reckwitz, 2012, p. 12),[1] with universally valid properties whose attribution to this entity is immutable. Over time, discussion around the subject diversified and made inroads into fields beyond philosophy; in discourses from political science and economics, for example, the concept of the subject is a key locus of divergent perspectives on forms of government, types of state and designs for life. The subject is a concept closely entangled with prototypical notions of citizenship and a citizen class and their shifts as time has passed. We have seen the successive prevailment of the idea that the modern subject is a "product of specific socio-cultural conditions" (Reckwitz, 2006, S. 9) whose definition would be incomplete without consideration of the factors set out here.

In recent times, trendsetting developments in theories of the subject have issued particularly from poststructuralist discourses. A highly influential milestone appears in the work of Michel Foucault, which has foregrounded the historically contextualised process via which a subject comes to be and regards the state of being a subject as secondary hereto. Foucault's oft-cited dictum of the 'death of the subject', alongside its multiple revisions and reinterpretations, marks an important point of reference for poststructuralist thought, notwithstanding the various controversies that have sprung up around it. Foucault's later work advanced the idea of specific 'Technologies of the Self' (Foucault, 1988), which, in their interaction with societal forms of knowledge and power relations, form the irreducible fundament of any historical analysis of contingent modes of being (Foucault, 2011, p. 9). Against this backdrop, Foucault explored the various cultural forms in which people develop knowledge of themselves and the associated disciplines of study, whose 'truth-games', interlinked with particular techniques, represent the point of departure for our self-knowledge (Foucault, 1988, p. 17). In the modern age, Foucault describes this form of *knowledge of self* as having ousted *care of self* from its former primacy: "In Greco-Roman culture knowledge of oneself appeared as the consequence of taking care of yourself. In the modern world, knowledge of oneself constitutes the fundamental principle" (ibid., p. 22).

One of the prominent names engaging with Foucault's work and elaborating its implications has been Judith Butler, best known for her work on gender theory, whose *The Psychic Life of Power* (Butler, 1997) fuses Foucault's concept of subjectivation with that of interpellation as proponed by the French philosopher Louis Althusser. Following in Foucault's footsteps, Butler proceeds from the assumption that people *are* not subjects a priori, but become or, more precisely, are made into them, and that it is these performative processes of subjection/subjectivation, drawing their shape from discourses and normative values, which merit analytical centrality. Becoming a subject appears, in this perspective, as an ambivalent process simultaneously of subjection/subjugation and of action after one's own mind. A frequently quoted excerpt from Butler's work refers to an exemplary situation she cites from Althusser, in which a policeman, with the words 'Hey, you there!', hails a pedestrian who, turning in response, becomes a subject through and by a combination of this hailing and that response of acknowledging the law and its validity—an acknowledgement encompassing the possibility that the subject might invoke it to empower him- or herself (ibid., p. 106). This point of view conceives of social orders and subjectivities as co-constitutive, not pre-existent, but requiring active production (there is more detailed discussion of this in the chapter by Britta Hoffarth in this volume).

Butler's work points up an aspect of discourses which is of substantive significance to the concern of this volume: their ubiquity and powerful effects both within and beyond their explicit reference in language: "discourses do not need to be explicitly cited in order to be deployed. Rather, multiple discourses are referenced through the meanings, associations and omissions embedded in the historicity of apparently simple and benign utterances and bodily practice" (Youdell, 2006. p. 514). Following this reading, we note the multi-faceted workings of discursivity, evident in language, yet also in embodied practices. I will return to this thought below, pursuing it further in light of material-medial manifestations of the discursive.

Numerous other authors have walked and extended the paths laid by Foucault and Butler. One focal point of current debates relates to concepts of distributed subjectivity (Alkemeyer et al., 2018), which read the subject as a collective entity rather than as one limited to a single individual and correspondingly examine not "subjectivation *in*, but *of* collectives" (Alkemeyer & Bröckling, 2018, p. 19). This emphasis on the collective element of subjectivation explores "how human [subjects], in their interplay with non-human entities, present themselves, assuming identity and readability, as this subject or that subject" (ibid., p. 24). This extension of the concept of subjectivation to encompass the co-action and inter-action of distributed entities directs our view both to large-scale social structures and to the significance of non-human entities to collective forms and means by which normative structures emerge, subjects locate themselves in or submit themselves to them, or rise up against hegemonic conditions in a spirit of emancipation. The approach taken by Rammert (2012) fits into this mould; while he does not give focal attention to the subject as such, his proposed concept of 'distributed agency' in socio-technical collectives emphasises the central moment of a distributed ability to act, opening up particularly towards human-technical interaction in various forms. All of this points the way towards an approach to subjectivation which leaves aside the familiar, individualistically-focused beaten track and, as I will discuss below, promises productive insights from its centring of collective processes under the assumption of a sociality constituted in and with the digital and media sphere.

1 Research on Subjectivation: An Outline

The increased currency of the subject as a concept in discourse, and the inspirations and variations on existing approaches it has engendered, shine a light on the analytical potential this concept contains, which we have been able only to touch upon briefly here. We note a mounting tendency to conceive of accesses to theories of the subject as proceeding beyond purely abstract, conceptual approaches, linking up with specific societal phenomena and finding a home in empirical methodologies. The overarching question in this direction of research, with its point of anchor in the theoretical presumption of subjectivation, seeks above all to illuminate the ways in which "actors relate to various different orders that structure an ideal way of being a subject" (Geimer et al., 2019, p. 3). The intent here is to overemphasise neither the power of the 'social orders that subjectivate'—for to do so would imply the marginalisation of thinking, active subjects—nor the autonomy these subjects enjoy. Instead, current research into subjectivation aims to examine the relationship between the subject positions on offer to subjects and the specific practices undertaken in this context, to the end of identifying whether subjects accept or reject these options for subjectivation, how they do so, the significance the actors involved attach to these acts, and the resistance they may offer instead, or the alternative proposals for subjectivation they may put forward (ibid., p. 4).

The subject, as a concept, has a notable role in various brands of discourse analysis; be they critical (Fairclough, 2010), poststructuralist (Angermuller, 2014), or informed by the sociology of knowledge (Keller et al., 2018), they all share some degree of referentiality to Foucault, and despite all differences they show in particular points, the subject, or perhaps rather subjectivation, is one (among several) of the overarching theoretical anchors of their methodological considerations, of which the reciprocal interaction of knowledge, power and discourse forms the key target. It is pertinent here, in regard to my proposed view of subjectivation research, to note that approaches from discourse analysis frequently stand accused of neglecting the actor's perspective and concomitantly overrating the significance of discursive structures. When viewed with subjectivation foremost in our minds, we will find it problematically one-sided to posit such a radical autonomy of discourses as effectively free-floating, severed from the daily realities of individuals. We might look, in this context, to build empirical bridges via subject analysis, with an associated, more distinct foregrounding of social practice. In a spirit that advances a notion of "subjectivation within social practices" (Alkemeyer, 2013), this strand of the discourse identifies an expansion of praxeological approaches to embrace theories of subjectivation as holding potential for multi-faceted insights into the complex processes by which sociality is produced and reproduced. The ultimate aim here is to initiate an approach to this research whose eyes are open to the ways in which, "commencing in the modern age, entities perceived as subjects, to whom we can attribute specific capacities to know, understand, act, judge, evaluate and reflect, come into being in historically mutable sets of social practices" (ibid., p. 36). The creation of a praxeological basis for subjectivation theory opens up to us a broad repertoire of methodological points of reference ranging from ethnographic approaches (Breidenstein et al., 2015) and artefact analysis (Lueger & Froschauer, 2018) to cartographic techniques (Both, 2015).

A related approach, drawing more strongly on discourse analysis in its poststructuralist sense, yet likewise illuminating connections between subjectivation and practices, is observable in the work of Wrana (2012a, b, 2015a) and of Wrana and Langer (2007) around the analysis of discursive practices. The greater centrality of discourse here does not detract from the authors' evident interest in matters of subjectivation. In a way, analysing discursive practices highlights the performative side of discourses, that is, their productivity via action and the completion of action. Wrana (2015b, p. 121) sets out a vision of discourse in this context thus: "I wish [here] to pursue a perspective that conceives of discourse not as originating and unfolding in the reproduction and stabilisation of the ordering structures of meaning, but as the operative and situated practice of structuring and ordering meaning". The analysis of discursive practices, then, would centre "contextualised acts of expression" (ibid.), with specific regard to their subjectivating function; rather than examining primarily *what* is spoken or subjectively 'meant', it would explore "how ensembles of meanings and objects are constructed" (ibid., p. 135). Alongside this stand numerous further approaches to the empirical study of subjectivation, which I cannot detail here without veering significantly off course. The key point here is that research on subjectivation accesses the entire range of qualitative methodologies and additionally generates highly diverse, theoretically rich analytical approaches in its endeavour to adequately meet its subject's complexity. It is doubtless the case that this great methodological flexibility is one of the strengths of this field, due in part to its capacity to provide tools distinctly apposite to the task; it is, however, simultaneously a weakness in that the empirical approach calls, as a rule, for extensive methodological modifications and acts of mapping the territory.

The particular appeal of the analysis of subjectivation to education science resides in the awareness it can raise of the configurations of power that rear their heads in all processes of pedagogy, in the form both of supra-individual contexts and of the "micro-physics of power" (Foucault, 1978, p. 26) that, in diverse ways, infiltrate and pre-mutate situations of teaching and learning, educational institutions, and processes of socialisation. This perspective directs a new light on pedagogical challenges such as the antinomies and ambivalences inherent in the act of educating and the paradoxical relationship of freedom and compulsion that characterises it. Key significance in this regard accrues to the matter of whether we conceive of a subject as 'weak' or 'strong'—the response here determines whether, where, and how we deem pedagogical interventions as appropriate or necessary; the extent to which such interventions empower or disempower; and the workings of the production and reproduction of normativity they occasion. This approach thus entails an interrogation of the role of education itself—as an academic discipline and a sphere of practical action—and the values it carries, implicit and otherwise (Färber, 2019). This investigation also uncovers governmental aspects of education and pedagogy (Weber & Mauer, 2006) and opens up a field for critique of practices within education due to its laying bare of configurations of power and knowledge and their implications for directive interaction with people of whom one is 'in charge'. The associated analysis of the relationship between technologies of power and technologies of the self directs our awareness sharply towards the difficult matter of laying down normative propositions in pedagogical practice. Butler's work on gender (Butler, 1990) is of particular relevance here, having resonated vigorously in education science and providing an outline for a continuous and ongoing engagement with the issues raised in theories of recognition and performativity (Jergus, 2012). Studies of the ordering structures at work in pedagogy likewise borrow from research on subjectivation, as is evident in areas such as the analysis of cultures of learning (Kolbe et al., 2008; Fritzsche et al., 2011). Focal exploration of the autonomy and heteronomy of acts of addressing and interpellation makes the normativity (or otherwise) of pedagogical processes amenable to description. In this way, working on an empirical basis, we gain access to a form of non-reductionist didactic decision-making which takes account of the multi-layered nature of the processes and practices at work in the school setting.

2 The Medial and Material Side of Subjectivation: Mapping the Lacunae in Education Science

Faced with such research potential, we might be forgiven for a degree of puzzlement at the rather hesitant response made by the discipline of education science to the debates around the changing concept of the subject which we have seen arise in the recent past. It is of use in this context to explore this conceptual shift more closely. Two of its main drivers, themselves reciprocally linked, are of significance here. One appears in the form of the material-cultural turn, whose challenge to existing positions in theories of the subject—via, for example, 'new materialism' (Dolphijn & van der Tuin, 2012; Gamble et al., 2019; Kissmann & van Loon, 2019)—consists in its radical conception of anthropocentrically defined categories such as thought and action as situated phenomena whose capacity to materialise in the first place is contingent upon specific configurations of human and non-human variables. Accordingly, the associated conceptualisation of the subject diverges from more classical variants which often equate 'the subject' with the human individual, relegate materiality to the status of a subsidiary contextual factor, and emphasise the subject/object dichotomy—a demarcation line which new materialism calls into question, taking the poststructuralist notion of the decentred subject to a new level and adding an "emphasis on the specific eventness and potentiality of material" (Folkers, 2013, p. 17) that rejects existing theories of the subject for, above all, their anthropocentrism. These approaches regard subjectivity per se as a hybrid, processual phenomenon involving various human and non-human factors.

The other catalyst of the reconceptualisations currently in process around the subject is related to the observation, in our recent past, of caesuras and changes in media cultures associated with the high importance of digital technologies in social interaction (Cheney-Lippold, 2019; Seyfert & Roberge, 2017; Stalder, 2018). Commentators on this phenomenon have repeatedly cited the overarching property of digital media, as manifest in their specific intertwinement of hardware and software, their networked data infrastructure and their underlying algorithmic processes, as a constitutive anthropological dimension of our age (Jörissen, 2014). The concept of (digital) 'mediality' has emerged in this context as a response to various developments. One of them is the socio-cultural shift, empirically in evidence over recent years, as a concomitant to the ubiquity of digital technologies. We do not need to look far to note the explosion in ways and means of access to media in the typical household, be it via mobile end devices, networks connecting them, time spent on digital media and platforms right across the socio-demographic spectrum; these unquestionable indicators of the transformation in our media use underline the need for a response from the academic disciplines of social sciences, cultural studies and, indeed, education science. Mediality, as a concept, frames attempts to note the changes these developments have wrought on aspects of our daily lives, or, put differently, to uncover the links between changes in mediality and changes in lifeworlds. It also signals a redirection of the view taken by theories of media, a more markedly operative perspective seeking to find out 'what media do when they are doing', how they co-produce worlds—and increasingly autonomously so, in the case of algorithm-based digital systems—and what changes specific forms of mediality engender in processes that bring meaning into being (Bettinger, 2020; Jäger, 2015).

In line with the generativist approach (Krämer, 2004) which holds that the medium is the message (McLuhan & Fiore, 1967), mediality, in our current world, means that media transcend the neutral, transmissive role the term suggests, and instead "create sense by transferring it. The transferring activity does not leave the transferred elements unchanged since it is carrying out a specific operation of embodiment by giving it a form or by creating a phenomenon" (Balke & Scholz, 2010, p. 40). Taking this view entails the assumption that forms of subjectivation, in our current world, take place through and by media, alongside their more analogue manifestations, and that this type of mediatedness, in the case of digitally networked technologies, is significantly different from other types. Krotz and Hepp (Hepp, 2013; Hepp & Krotz, 2014), for instance, posit the proliferation of mediality observable in our times and referenced in the umbrella term of 'digitality' as not amenable to explanation via a logic possessed of universal validity, but rather as rooted in multi-faceted influencing forces that, acting in various arenas of life and emerging via specific entanglements with the diverse socio-cultural conditions they meet, give birth to distinct mediatised worlds. We here face the difficulty posed by the fact that these media-shaped forces largely operate in the background. People's practices with media are mostly routine, habituated acts firmly embedded in our daily lives, acts upon which we do not consciously reflect, which, so to speak, 'disappear' even as we execute them (Krämer, 2008, p. 28) and, especially where they engage with digitally networked media, effectively mask their specific workings. Digital mediality, as an interaction of hardware, software and runtime (Passoth, 2017), takes place as an internally coordinated, background process on the basis of thoroughly enacted design choices aiming in many instances to generate a specific user experience and offering a menu of options for relationing. It is in this light that Jörissen (2015, p. 216) observes that, "in the context of the practices of daily life, things and thing-environments represent offers to engage in subjectivation and specifically offers to become user-subjects (consumers, prosumers, audiences, etc.) in particular ways". In many cases, it is the occurrence of malfunctions or failures that casts us into conscious awareness of this type of mediatedness; our disrupted experience may hold in our faces the agency of media's configurations in socio-medial structures, give us a brief glance at their power-shaped quality and lift the veil on the invitation to subjectivation at their core. It is not far from here to the notion of anthropomediality (Engell et al., 2013; Voss, 2010; Voss et al., 2019), which implies a concept of the subject that departs from the existentialism of Sartre, the transcendental philosophy of Kant and the dualistic subject/object distinction of Descartes—all of which posit the subject as an entity with the capacity for reflection on itself—and instead define it as primarily produced by the socio-cultural conditions around it, which in the context of this argumentation means digital mediality. The parallels to new materialism as outlined above are evident, and we accordingly note the existence of research that links mediality to materiality (van den Boomen et al., 2009; Thielmann & Schüttpelz, 2013; Spöhrer & Ochsner, 2017).

One of the current empirical challenges facing any analysis of subjectivation which strives to take appropriate account of mediality—in education science and beyond—is the matter of including digital data in analysis. While connections to artefact analysis readily reveal themselves at this juncture, we would do well not to lose sight of the necessarily selective nature of such a process, which can only ever cover a segment of the terrain. If, for example, we regard datafication as a central feature of digital mediality, we will indeed find ourselves prompted to incorporate large and diverse corpora of existing data, and then to wonder whether and, if so, how the quantitative approach this implies can fit with the primarily qualitative basis of subjectivation research to date. I see great potential merit in a productive dialogue between work on subjectivation and data science as to how each field might support and enrich the other. Further potential leads may stem from innovative work in cultural analysis which, to name an example, uses a variety of methods of visualisation (Manovic, 2020). In addition to this, a synthesis between the analysis of subjectivation and the discipline of critical software studies would appear promising (Jörissen & Verständig, 2017). Taking existing research approaches forward in the spirit outlined here would expand our horizon for comprehending the complexity and the power dynamics underlying processes of the generation of meaning as entanglements between the social and the digital/medial spaces.

I return now to my observation, made at the beginning of this section, that attempts in the discipline of education science to examine the phenomena of learning, education or socialisation have shown distinct reticence towards engaging theoretically or empirically with current innovations at the point of intersection between subjectivation and media research. A classical, dualistic conception of the subject, drawing on Enlightenment tradition, persists within the discipline. Work that does reference new developments in theories of the subject frequently fails to take account of mediality; discussions of mediality will often omit examination of the subject. An exemplary survey of discourses around theories of learning—very much education science's home turf—shows them as often markedly psychological in tendency, led, for example, by cognitivist approaches and lacking any reflection on the assumptions undergirding their theories of the subject (Künkler, 2011). There have been moves towards giving greater room to contextual factors, in work, for instance, on situated learning (Lave & Wenger, 1991); this said, the strong push in the last few years towards a focus on 'competencies' emphatically suggests an idea of learning as targeted towards the attainment of a specific output and/or the acquisition of 'skills' by and pertaining to the individual, with little attention to matters that would be the key concern of a subjectivation-oriented analysis. Theories of learning frequently reduce media to their instrumental capacity, and acknowledgement of their full socio-cultural implications may fall victim to a research preoccupation with what can be measured and quantified. Research into socialisation has also engaged very little with recent work on subjectivation (Grusec & Hastings, 2014; Hurrelmann et al., 2015). This is a field which traditionally pays less regard to the subject than to the individual as a socialised entity, the emergence of a personality and the acquisition of the capacity to take part in societal processes; theories of socialisation tend to assume that such processes of identity formation and personality development are finite, or at least pretty much so, concluding with the attainment of specific stages of development, with active and—in principle—self-determined involvement on the part of the subject (Färber, 2019, p. 81). The increasing awareness in this field of media as variables with relevance to and influence on socialisation has yet, in part, to progress beyond isolationist conceptions of 'the media' as distinct socialisation factors alongside with others.

Having identified and mapped these lacunae, I present this book as an attempt to fill in the gaps and cast new light on the complex, multi-layered interrelationships between subjectivation and digital mediality from an education studies perspective. Each chapter approaches a distinct facet of this area of interaction. I hope both to inspire new theoretical innovations in education science and to raise awareness of unbeaten methodological tracks towards research into subjectivation in medial-material configurations.

3 Overview of Chapters

The chapter by *Sabrina Schenk* explores the nature of the change in subjectivity in the digital context. It begins with analysis of work by Felix Stalder and Karin Knorr Cetina, informed by the social sciences and cultural studies, with the economy of knowledge (Stalder) and the postsocial (Knorr Cetina) as key points of reference. It is on this basis that Schenk identifies the necessity of conceptualising subjectivity in terms of network logics, and proceeds to attempt this endeavour via a comparative survey of three theoretical perspectives that share a critical positioning and an acknowledgement of practices mediated via technical or, put differently, material means. Schenk concludes her chapter by noting the paradoxes and the potential of such conceptualisations for education studies' engagement with ideas of 'networked subjectivity'.

Valentin Dander and Felicitas Macgilchrist present an examination of the field of intersection between datafication and civic education, with an emphasis on the analysis of the political subjectivities that emerge in the context of 'digital citizenship' and 'open data'. Working in alignment with critical discourse analysis and taking theories of political subjectivation as their starting point, the authors study various materials associated with 'School of Data', an initiative running workshops, aimed primarily at NGOs, on data, technologies and their status and impact in civil society. Their analysis identifies various forms of political subjectivity and leads them to an assessment of the role of data literacy in project-based educational practices; concluding, Dander and Macgilchrist pinpoint the need for a more explicit turn towards specific data-related practices if the concept of data literacy is to gain more appropriate and comprehensive foundations than have been in place to date.

The approach taken by *Britta Hoffarth* in her chapter draws on theories of media and *Bildung* in its analysis of subjectivation in the exemplary context of sexist hate speech. Referencing theoretical propositions by Foucault, Butler, Althusser and Adorno, Hoffarth turns a spotlight on the political dimension of performative conceptions of subjectivation. Following the theory of media advanced by Dieter Mersch, she defines mediality as "a fundamental moment of thinking and speaking", then proceeding to make visible "the specific conditions of becoming a subject—in relation to gender orders—[...] by empirically operationalising the dialectical relationship between self-techniques, invocations/addresses and figurations of knowledge via the concept of addressing". The conclusions for media education which she draws from her observations on internet violence with sexist motives point to the necessity of including an awareness of power relations in pedagogical thought.

Viktoria Flasche's contribution to the volume illuminates the interrelationship between the software architecture underlying social media platforms and young people's contemporary media practices. Working from a transactional perspective and drawing on Facebook, Instagram and TikTok as examples, Flasche uncovers the long-term processes of subjectivation in evidence in the entanglement between "the socio-technical structure of the social media platform in question and the individual's relationship with itself as reflected in multimedia representations of that self". In so doing, she demonstrates the resistance young people offer to the orders of visibility pushed by social media platforms, noting how they disrupt habitual structures of response via "aesthetic-tentative" practices, and highlights the potential for educative moments that emerges from these disruptions.

In my own chapter, I investigate possibilities for forging methodological links between discourse analysis and biographical research as a foundational basis for examining processes of *Bildung* in a context of digital mediality. Having noted and formulated the reciprocal productivity of these two research directions, and its particular manifestation in relation to the concept of the subject, I propose a methodological framework for analysing processes of *Bildung* as transformations in socio-medial configurations. My intent in so doing is to support a research perspective which seeks to release research on *Bildung* from a sole dependency on verbal biographical articulations and offer instead a conception of *Bildung*'s explicitly hybrid constitution, which in turn suggests the inclusion of digital artefacts in the methodology thus formulated. In this context, I conclude by drawing on extant research considerations around the analysis of material-discursive practices and exploring the potential for biographical research inherent in this approach.

I would like to express my thanks to the authors of this volume's chapters and to all others without whose contributions this volume would not have been possible. Alongside Katherine Ebisch-Burton's conscientious, observant and knowledgeable editing and translation work, I owe much to the encouragement and patience of Rebecca Wyde from Palgrave and Wendy Anne Kopisch of the Georg Eckert Institute for International Textbook Research. My thanks are also due to my student assistant, Saskia Draheim, for her support with formatting the chapters.

Notes

1. German quotes translated by Katherine Ebisch-Burton and Patrick Bettinger.

References

Alkemeyer, T. (2013). Subjektivierung in sozialen Praktiken. Umrisse einer praxeologischen Analytik. In T. Alkemeyer, G. Budde, & D. Freist (Eds.), *Selbst-Bildungen. Soziale und kulturelle Praktiken der Subjektivierung* (pp. 33–68). Transcript.

Alkemeyer, T., & Bröckling, U. (2018). Jenseits des Individuums. Zur Subjektivierung kollektiver Subjekte. Ein Forschungsprogramm. In T. Alkemeyer, U. Bröckling, & T. Peter (Eds.), *Jenseits der Person. Zur Subjektivierung von Kollektiven* (pp. 17–31). Transcript.

Alkemeyer, T., Bröckling, U., & Peter, T. (Eds.). (2018). *Jenseits der Person. Zur Subjektivierung von Kollektiven.* Transcript.

Angermuller, J. (2014). *Poststructuralist discourse analysis. Subjectivity in enunciative pragmatics.* Palgrave Macmillan.

Balke, F., & Scholz, L. (2010). The medium as form. In L. Jäger, E. Linz, & I. Schneider (Eds.), *Media, culture, and mediality. New insights into the current state of research* (pp. 37–48). Transcript.

Beer, R. (2014). Das Subjekt im Wandel der Zeit. In T. Carstensen, C. Schachtner, H. Schelhowe, & R. Beer (Eds.), *Digitale Subjekte. Praktiken der Subjektivierung im Medienumbruch der Gegenwart* (pp. 215–272). Transcript.

Bettinger, P. (2020). Materialität und digitale Medialität in der erziehungswissenschaftlichen Medienforschung. Ein praxeologisch-diskursanalytisch perspektivierter Vermittlungsversuch. In S. Iske, J. Fromme, J. Bastian, M. Pietraß, & K. Rummler (Eds.), *Jahrbuch Medienpädagogik 15: Erziehungswissenschaftliche und medienpädagogische Online-Forschung: Herausforderungen und Perspektiven* (pp. 53–78). Zeitschrift MedienPädagogik.

Both, G. (2015). Praktiken kartographieren. Was bringt Clarkes Situational Analysis für Praxeografien? In F. Schäfer, A. Daniel, & F. Hillebrandt (Eds.), *Methoden einer Soziologie der Praxis* (pp. 197–214). Transcript.

Breidenstein, G., Hirschauer, S., Kalthoff, H., & Nieswand, B. (2015). *Ethnografie. Die Praxis der Feldforschung* (2nd ed.). UVK.

Butler, J. (1990). *Gender trouble: Feminism and the subversion of identity.* Routledge.

Butler, J. (1997). *The psychic life of power. Theories in subjection.* Stanford University Press.

Cheney-Lippold, J. (2019). *We are data. Algorithms an the making of our digital selves.* New York University Press.

Dolphijn, R., & van der Tuin, I. (2012). *New materialism: Interviews & cartographies.* Open Humanities Press.

Engell, L., Hartmann, F., & Voss, C. (Eds.). (2013). *Körper des Denkens. Neue Positionen der Medienphilosophie.* Wilhelm Fink.

Fairclough, N. (2010). *Critical diskcourse analysis. The critical study of language* (2nd ed.). Routledge.

Färber, C. (2019). Subjektivierung in der Pädagogik. Das Subjekt zwischen Ent- und Ermächtigung. In N. Ricken, R. Casale, & C. Thompson (Eds.), *Subjektivierung. Erziehungswissenschaftliche Theorieperspektiven* (pp. 75–92). Beltz Juventa.

Folkers, A. (2013). Was ist neu am neuen Materialismus?—Von der Praxis zum Ereignis. In T. Goll, D. Keil, & T. Telios (Eds.), *Critical Matter. Diskussionen eines neuen Materialismus* (pp. 17–34). Edition Assemblage.

Foucault, M. (1978). *Discipline and punish. The birth of the prison.* Random House.

Foucault, M. (1988). Technologies of the self. In L. M. Martin, H. Gutman, & P. H. Hutton (Eds.), *Technologies of the self. A seminar with Michel Foucault.* Tavistock Publications.

Foucault, M. (2011). *The courage of the truth (The government of self and others II) Lectures at the Collège de France 1983–1984.* Palgrace Macmillan.

Fritzsche, B., Idel, T.-S., & Rabenstein, K. (2011). Ordnungsbildung in pädagogischen Praktiken. Praxistheoretische Überlegungen zur Konstitution und Beobachtung von Lernkulturen. *ZSE Zeitschrift für Soziologie der Erziehung und Sozialisation, 1*, 28–43.

Gamble, C. N., Hanan, J. S., & Nail, T. (2019). What is new materialism? *Angelaki. Journal of the Theoretical Humanities, 24*(6), 111–134. https://doi.org/10.1080/0969725X.2019.1684704

Geimer, A., Amling, S., & Bosančić, S. A. (2019). Einleitung: Anliegen und Konturen der Subjektivierungsforschung. In A. Geimer, S. Amling, &

S. A. Bosančić (Eds.), *Subjekt und Subjektivierung. Empirische und theoretische Perspektiven auf Subjektivierungsprozesse* (pp. 1–15). Springer VS.

Grusec, J. E., & Hastings, P. D. (Eds.). (2014). *Handbook of socialization. Theory and research* (2nd ed.). Guilford Publications.

Hepp, A. (2013). The communicative figurations of mediatized worlds: Mediatization research in times of the 'mediation of everything'. *European Journal of Communication, 28*, 615–629. https://doi.org/10.1177/0267323113501148

Hepp, A., & Krotz, F. (Eds.). (2014). *Mediatized worlds. Culture and society in a media age*. Palgrave Macmillan.

Hurrelmann, K., Bauer, U., Grundmann, M., & Walper, S. (Eds.). (2015). *Handbuch Sozialisationsforschung* (8th ed.). Beltz.

Jäger, L. (2015). Medialität. In E. Felder & A. Gardt (Eds.), *Handbuch Sprache und Wissen*. de Gruyter.

Jergus, K. (2012). Politiken der Identität und der Differenz. Rezeptionslinien Judith Butlers im erziehungswissenschaftlichen Terrain. In N. Ricken & N. Balzer (Eds.), *Judith Butler: Pädagogische Lektüren* (pp. 29–53). Springer VS.

Jörissen, B. (2014). Digitale Medialität. In C. Wulf & J. Zirfas (Eds.), *Handbuch Pädagogische Anthropologie* (pp. 503–513). Springer.

Jörissen, B. (2015). Bildung der Dinge: Design und Subjektivation. In B. Jörissen & T. Meyer (Eds.), *Subjekt Medium Bildung* (pp. 215–233). Springer VS.

Jörissen, B., & Verständig, D. (2017). Code, Software und Subjekt. Zur Relevanz der Critical Software Studies für ein nicht-reduktionistisches Verständnis "digitaler Bildung". In R. Biermann & D. Verständig (Eds.), *Das umkämpfte Netz. Macht- und medienbildungs- theoretische Analysen zum Digitalen* (pp. 37–50). Springer VS.

Keller, R., Hornidge, A.-K., & Schünemann, W. J. (Eds.). (2018). *The sociology of knowledge approach to discourse. Investigating the politics of knowledge and meaning-making*. Routledge.

Kissmann, U. T., & van Loon, J. (Eds.). (2019). *Discussing new materialism. Methodological implications for the study of materialities*. Springer VS.

Kolbe, F.-U., Reh, S. R., Fritzsche, B., Idel, T.-S., & Rabenstein, K. (2008). Lernkultur: Überlegungen zu einer kulturwissenschaftlichen Grundlegung qualitativer Unterrichtsforschung. *Zeitschrift für Erziehungswissenschaft, 11*(1), 125–143.

Krämer, S. (2004). Was haben 'Performativität' und 'Medialität' miteinander zu tun? Plädoyer für eine in der 'Aisthetisierung' gründende Konzeption des Performativen. In S. Krämer (Ed.), *Medialität und Performativität* (pp. 13–32). W. Fink.

Krämer, S. (2008). *Medium, Bote, Übertragung. Kleine Metaphysik der Medialität*. Suhrkamp.

Künkler, T. (2011). *Lernen in Beziehung. Zum Verhältnis von Subjektivität und Relationalität in Lernprozessen*. Transcript.

Lave, J., & Wenger, E. (1991). *Situated learning. Legitimate peripheral participation*. Cambridge University Press.

Lueger, M., & Froschauer, U. (2018). *Artefaktanalyse. Grundlagen und Verfahren*. Springer VS.

Manovic, L. (2020). *Cultural analytics*. MIT Press.

McLuhan, M., & Fiore, Q. (1967). *The medium is the massage*. Penguin Books.

Passoth, J.-H. (2017). Hardware, software, runtime. Das Politische der (zumindest) dreifachen Materialität des Digitalen. *Behemoth, 10*(1), 57–73.

Rammert, W. (2012). Distributed agency and advanced technology. Or: How to analyse constellations of collective inter-agency. In J.-H. Passoth, B. Peuker, & M. Schillmeier (Eds.), *Agency without actors? New approaches to collective action* (pp. 89–112). Routledge.

Reckwitz, A. (2006). *Das hybride Subjekt. Eine Theorie der Subjektkulturen von der bürgerlichen Moderne zur Postmoderne*. Velbrück Wissenschaft.

Reckwitz, A. (2012). *Subjekt* (3rd ed.). Transcript.

Seyfert, R., & Roberge, J. (Eds.). (2017). *Algorithmic cultures. Essays on meaning, performance and new technologies*. Routledge.

Spöhrer, M., & Ochsner, B. (Eds.). (2017). *Applying the actor-network theory in media studies*. IGI Global.

Stalder, F. (2018). *The digital condition*. Polity Press.

Thielmann, T., & Schüttpelz, E. (Eds.). (2013). *Akteur-Medien-Theorie*. Transcript.

van den Boomen, M., Lammes, S., Lehmann, A.-S., Raessens, J., & Schäfer, M. T. (Eds.). (2009). *Digital material. Tracing new media in everyday life and technology*. Amsterdam University Press.

Voss, C. (2010). Auf dem Weg zu einer Medienphilosophie anthropomedialer Relationen. *Zeitschrift für Medien- und Kulturforschung (ZMK), 2*, 169–184.

Voss, C., Krtilova, K., & Engell, L. (Eds.). (2019). *Medienanthropologische Szenen. Die conditio humana im Zeitalter der Medien*. Wilhelm Fink.

Weber, S., & Mauer, S. (Eds.). (2006). *Gouvernementalität und Erziehungswissenschaft. Wissen—Macht—Transformation*. Sprunger VS.

Wrana, D. (2012a). Diesseits von Diskursen und Praktiken. Methodologische Bemerkungen zu einem Verhältnis. In B. Friebertshäuser, H. Kelle, H. Boller, S. Bollig, C. Huf, A. Langer, M. Ott, & S. Richter (Eds.), *Feld und Theorie. Herausforderungen erziehungswissenschaftlicher Ethnographie* (pp. 185–200). Opladen, Berlin & Toronto.

Wrana, D. (2012b). Theoretische und methodologische Grundlagen der Analyse diskursiver Praktiken. In D. Wrana & C. M. Reinhard (Eds.), *Professionalisierung in Lernberatungsgesprächen. Theoretische Grundlegungen und Empirische Untersuchungen* (Vol. 3, pp. 195–214). Budrich.

Wrana, D. (2015a). Zur Analyse von Positionierungen in diskursiven Praktiken. Methodologische Reflexionen anhand von zwei Studien. In S. Fegter, F. Kessl, A. Langer, M. Ott, D. Rothe, & D. Wrana (Eds.), *Erziehungswissenschaftliche Diskursforschung. Empirische Analysen zu Bildungs- und Erziehungsverhältnissen* (pp. 123–141). Springer VS.

Wrana, D. (2015b). Zur Methodik einer Analyse diskursiver Praktiken. In F. Schäfer, A. Daniel, & F. Hillebrandt (Eds.), *Methoden einer Soziologie der Praxis* (pp. 121–143). Transcript.

Wrana, D., & Langer, A. (2007). An den Rändern der Diskurse. Jenseits der Unterscheidung diskursiver und nicht-diskursiver Praktiken. *Forum Qualitative Sozialforschung, 8*.

Youdell, D. (2006). Subjectivation and performative politics—Butler thinking Althusser and Foucault: Intelligibility, agency and the raced-nationed-religioned subjects of education. *British Journal of Sociology of Education, 27*(4), 511–528.

CHAPTER 2

'Network Subjectivity' in the Digital Condition: Three Theoretical Envisionings

Sabrina Schenk

Abstract There are currently various academic accounts of contemporary sociality. While the concept of the 'digtital condition' seems to lend itself to a cultural interpretation, those of 'informationalism' and 'postsociality' may be appropriate for describing, from a sociological perspective, specific phenomena of changes to sociality in the digital condition. The emergence of the term 'knowledge economy' appears to invite an intersection of cultural and sociological perspectives and proceeding from this starting point, this chapter identifies the shared parameters of these perspectives in relation to the conditions of subjectivity, of high relevance to education, in a networked, media-driven society. The work underlying the chapter

All quotations in the text that are taken from German-language sources have been translated by the author. I would like to thank the editor Patrick Bettinger for his intellectual support and particularly the proofreader Katherine Ebisch-Burton for her empathy with and commitment to this text.

S. Schenk (✉)
Institut für Allgemeine Erziehungswissenschaft (WEI), Goethe-Universität Frankfurt, Frankfurt am Main, Germany
e-mail: sabrina.schenk@em.uni-frankfurt.de

P. Bettinger (ed.), *Educational Perspectives on Mediality and Subjectivation*, Palgrave Studies in Educational Media,
https://doi.org/10.1007/978-3-030-84343-4_2

examines these issues via the metaphor of 'network subjectivity', employing the findings of an empirical study, theory of government, and psychoanalysis.

Keywords Digitisation • Digital culture • Network society • Postsocial • Subjectivity

Post-structuralist approaches, particularly that of Michel Foucault, have become strongly influential in the exploration of socially mediated subjectivity in education studies within the last two decades. From this perspective, theories of subjectivity that rely on traditions influenced by transcendental philosophy or the philosophy of consciousness have come in for particular criticism (see, for example, the contributions in Cramer et al., 1990). More recent analyses of subjectivity tend to take a sociophilosophical or cultural starting point (cf. Reckwitz, 2008, pp. 18, 120). Losing its previous status as a transcendental, monadic-autonomous figure of thought with universal implications, the subject now appears as the result of power-shaped processes of subjectivation "under specific sociocultural conditions" (ibid., p. 10). This said, sociology and education studies have not abandoned the concepts of subjectivity and subjectivation, indeed revisiting them in multiple variations (see Alkemeyer et al., 2018). Following Ricken et al. (2019, p. 7), these concepts appear to possess the capacity to enable empirical and analytical researchers to swerve previously dominant dichotomies (such as that between active control and passive submission) and rethink processes of social and cultural genesis and hence the traditional concepts of education and *Bildung*. In this respect, the analysis of the "entrepreneurial self" (Bröckling, 2016) as one of the currently dominant forms of subjectivity has received great attention.

Rather than following Bröckling's focus on 'governmentality', with its identification of "programmes" (ibid., p. 12) and "regimes" (ibid., p. 13) of subjectification, this chapter seeks to pinpoint conceptions of subjectivity that are useful in the present age. In its exploration of ways of conceiving of and describing subjectivity in the digital condition, or, putting it differently, the digital culture, it notes the currently increasing use of the category of the 'network' in social theory and theories of subjectivation alike. I argue that this category could well prove productive in illuminating both the specifics of a particular sociality and one possible form of subjectivity within the digital condition.

I will commence my argument with two proposals on conceptualising society in the digital condition: the recent work by Felix Stalder from a cultural studies perspective, and the sociologically oriented account of the 'postsocial' put forward by Karin Knorr Cetina, which points towards some characteristics of an 'object-centered subjectivity'. The subsequent section of the chapter, proceeding from both theorists' use of the 'network' metaphor, will outline three approaches which also use this image to characterise the changes that subjectivity experiences within the digital condition.

1 The Digital Condition and the Network as the 'Social Morphology of the Present Age'

The terms 'digitalisation' and 'digitisation' currently serve to encompass sweeping social and societal changes and simultaneously reduce them, and the associated challenges, fundamentally to a technological common denominator. In response to a search query for the period 2015–2020 (made on 5 June 2020), Google Scholar lists approximately 4400 findings with the keyword 'digitalization' (1100 for 'digitalisation') in the title and around 45,600 (17,000) findings with this keyword in the text. 'Digitization' returns 3300 (630 for 'digitisation') results in titles and 52,400 (18,200 for 'digitisation') in texts. Most book titles—for both keywords—revolve around the changes to business and work brought about by digital information and communication technologies (ICT) or by the use of artificial intelligence (AI), i.e. learning algorithms. In the German-language discourse, 'digitalisation' or 'digitisation'—as the relatively hermetic concept *Digitalisierung*—has increasingly been appearing in the context of pedagogy and education studies, alongside other disciplines. The American Education Resources Information Center (ERIC), however, reports only 14 title hits for 'digitalization' (7 for 'digitalisation'), 55 for 'digitization' (15 for 'digitisation') and 77 (46) and 290 (55) keyword-based results respectively. This points to a distinct diversity of forms and collocations for the term in English. Indeed, changing the search term to 'digital' and rerunning the ERIC search returns approximately 6500 title- and 18,400 keyword-containing publications. The findings for the keywords "digital condition" and "digitality" contrast starkly, with very low hit rates (under 100) in both English- and German-language databases.[1] It is evident that these new discursive buzzwords cannot yet claim the same relevance in broader discourse as they have attained in academic disciplines such as media and cultural studies. *The Digital Condition* is a book, first published in English in 2018 and tackling its subject from a cultural studies perspective, by Felix Stalder,[2] who works "as a professor of digital culture and network theories at the Zurich University of the Arts and as an independent researcher/organizer with groups such as the Institute for New Cultural Technologies (t0) and the technopolitics group in Vienna" (http://felix.openflows.com/node/4). His account of the "digital condition" (i.e. digitality) distinguishes it as a cultural quality from "digitalization", which he associates with specific technological developments (such as the transformation of the materiality of objects by scanning; see Stalder, 2018, p. 61). The book defines the digital condition as a phenomenon that "has become quotidian and dominant. It forms a cultural constellation that determines all areas of life, and its characteristic features are clearly recognizable" (ibid., p. 57). Stalder dates the beginnings of digital culture back to nineteenth-century processes of social change (ibid., p. 41). This line of argument reverses the relevance of the technological processes of digitisation to current social changes, suggesting that it is not the invention of digital technologies or the expansion of the internet that has driven these transformations, but rather separate cultural developments that have begun to intertwine with these technologies. One of the book's fundamental theses is that cultural and social processes—as specific structures of knowledge and action—always precede technological processes and embed these within them.

Stalder pinpoints one of the digital condition's cultural and social sources in the erosion of patriarchal, heteronormative power relations in the course of the twentieth century's new social movements, exemplified in the development of the gay rights movement in the Federal Republic of Germany since 1969 (cf. ibid., p. 23). He further identifies post-colonialism, with its tendency towards the increasing fluidity of collective identities and towards an emphasis on hybridity, as a complement to and amplifier of the digital condition (cf. ibid., p. 33). Third, and most significantly in relation to sociological theories, Stalder traces the expansion of the knowledge economy beginning in the early 1950s, the concomitant emergence of the consumer society from the 1960s, and the rise of flexibility in labour and terms of employment in the 1970s (cf. ibid., p. 13). Ultimately, a close connection appeared between this flexibility and technological digitisation when, around the turn of the millennium, the internet became a mass medium and the creative industries evolved into a political agenda.

The upshot of this structural change in economics and labour, manifesting through the interaction of these three lines of cultural development, has been that the social basis of cultural production expanded and gained in heterogeneity, insofar as previously excluded social groups developed their own language and began to take their part in these discourses. Alongside this is the multiplication of those directly engaged in cultural production (cf. ibid., p. 12). This description of what constitutes digital culture, or the digital condition, must also encompass the effects wrought by the processes of culturalisation of the economy and of the economisation and technologisation of the world (cf. ibid., p. 35): Advertising and sales promotion imbues specific 'lifestyle' products with the alleged power to transform the consumer's living environment into a distinct "experiential world", and their slogans migrate into everyday communication as personal life mottos. Digitisation, networking and new communication technologies turn consumers into self-producers, self-designers. They also generate new practices and methods relating to "how people orient themselves and act in this changed informational environment", as Stalder points out in an interview for *First Monday* (https://firstmonday.org/ojs/index.php/fm/article/view/9409/7574). Referentiality, community and algorithmicity as three cultural forms characterising the digital condition simultaneously bring into being shared patterns of the creation of meaning and orientation, manifest in activities—such as tweeting and retweeting, sharing and liking—that channel the individual's attention by "filtering certain things out of the chaotic information sphere" (ibid.) and establish a communality, undergirded by prearranged algorithmic processes.

My reading of Stalder's analysis locates it primarily in the concept of the knowledge economy, in which it holds common ground with sociological studies; this leads to the emergence of points of contact with sociological accounts of contemporary society. Stalder's borrowings from sociological considerations presumably stem from his engagement with the work of the sociologist Manuel Castells (2010),[3] whose thesis of the "rise of the network society" Stalder takes up in *The Digital Condition*. Stalder (2018, p. 53) identifies one of the hallmarks of digital culture in the new space for networks created by digital technologies, an area located between the institutional or public space and the private or personal space, and constituted by digital communication. It is in this context that he explicitly endorses Castells' assertion that "[n]etworks constitute the new social morphology of our societies" (Castells, 2010, p. 500).[4]

2 Postsociality and Subjectivity Mediated via Objects

Alongside and antecedent to Stalder's analysis of contemporary digitality, sociological studies have also taken the knowledge economy as a starting point, following older discussions around the notion of a knowledge society. Karin Knorr Cetina, a sociologist of knowledge, uses the term of the "postsocial", fusing the concept's sociological dimension with an element drawing on social psychology, to characterise the contemporary changes in forms of society and sociality in globalised information societies. The "postsocial" signifies a sociality that no longer revolves around human beings and human interactions, but instead consists in the interconnection of a material, technological and informational dimension which embraces the partial replacement of humans by objects and generates an "object-centered sociality as a social form that constitutes something like the reverse side of the coin of the contemporary experience of individualization" (Knorr Cetina, 1997, p. 9). We see here that the starting point of her sociological description of subjectivity, drawing on the characteristics of the digital culture (Stalder) and the conditions of informationalism (as Castells put it), lies, as do Stalder's and Castells', in a social theory. In her outlines of social relations in a "postsocial knowledge society" (ibid., p. 25), Knorr Cetina (2005, p. 588) assumes that, to subjects, contemporary society manifests itself primarily as a mirror of a "media, image, and knowledge culture". This assertion encompasses at least two basic implications. First, it presupposes that there is value for the temporally specific analysis of society in theories of subject genesis such as that of the 'mirror stage' proposed by the psychoanalyst Jacques Lacan, who ascribes key significance to the initiation of a sense of the self as a distinct entity in the child's self-recognition in a mirror. Second, it discounts as invalid the idea of a society in which subjectivity is primarily constituted by (human) intersubjectivity, as in the role theory of the interactionist George Herbert Mead.[5] The act of refocusing implied in an acceptance of these premises brings a decisive shift into view as to what we regard as the preconditions of society; Knorr Cetina uses the term "postsocial" for this new dispensation, similarly noting and illuminating the associated changes in the preconditions of subjectivity. She observes that object worlds and technical infrastructures have expanded massively within the social sphere, which also means that "objects displace human beings as relationship partners and embedding environments, or […] increasingly mediate human relationships, making the latter dependent upon the former" (Knorr Cetina, 2005, p. 586). This analysis strikes a chord in light, *inter alia*, of the prevalence of digital technologies in our present-day communication, such as chatbots in customer services and, of course, the proliferation of email and messenger services in everyday communication across the globe.

Historically, this development has gone hand in hand with a rise in individualisation, which has precipitated the dissolution of previously dominant complex systems of social organisation in favour of networks incorporating the "socially unrelated members of the population" (ibid., p. 588). We therefore find ourselves required to identify new means of analysing these "alternative forms of binding self and other, changes in the structure of the self that accommodates these forms, and forms of social imagination that subordinate sociality to new promises and concerns" (ibid.), which, in the context of this chapter, means the structure and the promises of network subjectivity. Knorr Cetina finds in Lacan's mirror stage a model for an object-centred self that is mediated through objects or technologies rather than through human interaction. She places emphasis on the libidinal dimension of interacting with things and, above all, with what she terms "knowledge objects", ascribing a central role to the auto-affection of subjects in their interaction with things and objects, a phenomenon perhaps best understood via Lacan's model of the dynamics of lack and desire established in the mirror stage (cf. Knorr Cetina, 1997, p. 14).[6] The mirror stage describes the concept of desire, i.e. a permanent striving for wish fulfilment, in view of a constitutive lack or withdrawal of this fulfilment, in a manner similar to how the small child sees itself in the mirror as a whole, which clashes with its self-experience as incompetent. In adults, the 'perfect' other can be understood as a foil of the self's own desire. The constitutive lacuna between the other as the idealised self and the self's own lifelong experience of new imperfections manifests itself in a dynamic of desire: The construction and deconstruction of desired objects as soon as they are partially achieved generates a perpetual motion (see also Mayer & Schenk, 2018).

Knorr Cetina's harnessing of this model for the definition of a postsocial form of self-formation highlights the self's permanent motivation to expand its own emotional life and desire through auto-affection, rendering it suited to a society in which the "the mirror is exteriorized in a media, image, and knowledge culture", with its "professional image industries that project images and stage 'wholeness'" (Knorr Cetina, 2005, p. 588). Stalder has analysed this close intertwinement of subjects with material consumption and the culture industry as an interrelationship between the culturalisation of the economy and the economisation of culture. Along similar lines, Knorr Cetina defines the postsocial subject as part of a culture that "is centered on material, technological, and informational processes" (ibid., p. 590). In the object worlds of the culture industry, auto-affection becomes attached to objects outside the self and emotional bonds and relationships arise that overlap with, if not replace, social bonds and their significance.

In what follows, I will argue that the concept of the network encapsulates the conditions under which subjectivity comes to be in a postsocial digital culture. Knorr Cetina (2005, p. 587) asserts that in the postsocial phase, "complex organizations are dissolved into networks"; the declining significance of overarching social structures and relationships is in inverse proportion to the ascendancy of individuals' networking activities, carried by information and communication technologies (ICT) whose primary orientations are typically towards things or objects (such as other technologies) rather than people. She argues that particularly in contexts of knowledge-based work or in the structuring of everyday life via technology, interaction with these objects creates a relationality of reciprocal claims—to attention, focus and interaction, for example—between objects and subjects, and draws the subject in auto-affectively. Knorr Cetina (ibid., p. 589) proposes the term "interspecies reciprocity" as a point of access to this relationality.[7]

3 Subjectivity in the Digital Culture of the 'Network Society'

The rise of network theories corresponds to the increasing prominence of the phenomenon of the network "in people's everyday practices" (Jörissen, 2016, p. 231). The sociologist Manuel Castells (2010) was one of the early proponents of the term "network society", having devoted extensive analysis to the idea in 1996, at the beginning of the internet's emergence into public life. While Stalder sees the 'digital condition' as rooted in cybernetics, Castells (2010, p. 21) uses "informationalism" as the organising principle of his analysis of contemporary society, with the "network" appearing as a concept describing the results of his observations. Castells (ibid., pp. 3, 22) contrasts the network with the (individual and collective) self and therefore does not use this term for the analysis of subjectivity in the condition of informationalism (see also Nollmann, 2011, p. 638). One of the factors inhibiting a transfer of the concept of the network to the level of subjectivity is the fundamentally technical and material nature, within Castells' theoretical framework, of the "networking logic" (Castells, 2010, p. 52) embodied by the internet. From a sociological perspective, then, we can conceive of the network as a "structural feature of contemporary societies" (Holzer, 2006, p. 6). Further, the empirical continuations of Castells' proposal of the network as "a new dominant structure" (Winter, 2010, p. 29) emerging from cultural studies do not go as far as to apply this proposition to subjects themselves. Rainer Winter, for example, assumes that the decentralised structures of digital media are reflected in the network-like organisational form of social movements (ibid., p. 65), and not in the structures of networking subjectivities themselves. Consequently, he limits the range of the network metaphor to the sphere of collective dynamics, excluding forms of subjectivity from its reach.

By contrast with these apparent limitations, Castells (2010, p. 70) himself develops the "morphology of the network" (as defined by Kelly) from the idea of "the convergence between the evolutionary topology of living matter, the open-ended nature of an increasingly complex society, and the interactive logic of new information technologies". This evident capacity of the network to connect disparate phenomena might suggest to us that the concept's use in explorations of the formation of subjectivity in the digital condition, informationalism or postsociality is no random choice. In contrast to its function in social theory as a term in the critical analysis of our time, the network appears in explorations of subjectivation as more of a heuristic and metaphorical term.[8]

What follows will outline three such explorations on an exemplary basis. Their divergent theoretical contexts notwithstanding, they share a centring of technically or materially mediated practices, processes or infrastructures and a concern with their description rather than a primary focus on a large-scale framework of social theory such as the "network society". They also share a critical perspective, which appears explicitly in the observation by Paulitz (2005, pp. 11, 268) that both the instrumentalising and rationalising view of the network's *use* and the restrictive, overly pedagogically imbued view of its *effects* fall short in terms of understanding the interactions that produce a 'network subjectivity'. Raising the question of subjectivity in analogy to the network metaphor in the digital condition thus links both "the current formation of everyday culture in increasingly networked relationships" and the effects of that culture's transformation through "the shift in media worlds associated with digitalisation" (Hepp, 2010, p. 230).[9] Although none of these three approaches explicitly uses the term 'network subjectivity', I propose that it has the capacity to encompass shared features of all three, and therefore intend in what follows to put it forward as an umbrella term covering one possible form of subjectivity in the digital condition, to which the examples I will now discuss each contribute.[10]

Network Subjectivities as Intersections of Hyper-Linked "Nodes and Lines", both Multidimensional and Partial (Paulitz)

The sociologist Tanja Paulitz (2005, p. 28) proceeds from "open interpretations of the image of the net/network [*Netz*] as a diagnostic term for (current) processes of social transformation".[11] Her empirical study of two research projects in the fields of e-learning and Computer Supported Collaborative Work (CSCW) analysed documents, such as presentations, minutes and working papers, and expert interviews and produced findings to which she applies the collective designation "net/work subjectivity/ies [*Netzsubjektivität/en*]". Her use of *Netz* is explicitly as a "metaphor" (ibid., p. 23); she criticises what she perceives as its contemporary function as a "homogeneous principle for explaining the world" (ibid., p. 24) which suggests the possibility of direct access to subject areas which are supposed to be similar, but are in fact different and divergent. The pluralisation of "net/work subjectivities" seeks to resist this pull towards homogeneity.

From a methodological point of view, such "net/work subjectivities" are constructs drawn by Paulitz from her interpretations of the materials she analysed, which come about as the "result of a cultural and social practice (of designation)" (ibid., p. 13) on the part of the interviewees. In what follows, I will argue in favour of going beyond Paulitz' approach here by distinguishing two concepts: the eponymous 'net/work subjectivities' and the concept of 'networked subjectivity'. Paulitz' analysis of "net/work subjectivities" explicitly opposes a dualistic juxtaposition of subjectivity and the net/work as found in instrumental concepts of "net use" which often entail pedagogical discourses of opportunity and risk (see ibid., p. 11). It is in this rejection of dualism that I perceive the ultimately fruitful insight of her study, and I second this perspective in this chapter. Such instrumental discourses, however, are often everyday in nature, and do not necessarily appear in connection with an "enlightened concept of the subject" or a "model of repressive power" (ibid., p. 12), as Paulitz suspects them to do. In reinterpreting her material, I identify it partly as a collection of instances of "networked subjectivity", i.e. the dualism of subjectivity and the net/work, particularly when she elaborates on subjects in interaction with their virtual workspaces. Paulitz asserts that "actors who engage with media" (ibid., p. 245) bring their sensory perceptions, their "physical materiality and their personal complexity" (ibid., p. 246) into play against the virtual workspaces, even perceiving themselves as "exposed to the risks of control and functional appropriation" concomitant to this interaction (ibid.). Paulitz herself speaks of "networked subjectivity" (ibid., p. 245) in this context and adjoins it as one dimension of net/work subjectivities. Instead, I propose to link this concept with the juxtaposition of subjectivity and the network and distinguish it from the concept of "net/work subjectivities", despite Paulitz' evaluation of the two concepts as synonymous. The primary thrust of the idea of "net/work subjectivities", in my view, is towards an interwovenness of subjectivity and the net/work. Paulitz identifies this as taking place in the interviews she analyses via the interpretation of social and technical processes using the same descriptive categories, resulting in accounts, for instance, of the computer as a colleague. We witness the description of information technologies "in categories of the social" (ibid., p. 250); computers connected via the internet can, for example, represent a "'society' of heterogeneous actors who communicate with one another worldwide using common 'languages'" (ibid., p. 250). Analogously, their 'networking capability' (as a technical component) appears as social potential within communicative processes. The converse instance is that of depicting social processes in terms of informational structures, describing them, for instance, via a model of "nodes and lines"[12] or in accordance with the "pattern of hyperlink procedures on the web" (ibid., p. 255). With regard to the structural logic of "net/work subjectivities", Paulitz perceives "a reciprocal productivity between the refiguration of the technical as a social context and the refocusing of social interaction as a (technical) functional entity." (ibid., p. 265)

The notion of "nodes and lines" as a vision of cooperative collaboration in the virtual workspace takes on a central role (cf. ibid., p. 204). The platform emerges as a space in which, via meta-information, technical and non-technical actors (files, people, projects, messages) appear as "nodes" and in which hyperlinks enable their interactive and communicative interlinking or interrelation as "lines". This analysis identifies a multidimensional quality of network subjectivities: "Every element in the virtual workspace becomes multidimensional in principle, a multiplied 'node', which is linked to others multiple times and in multiple perspectives via 'lines'" (ibid., p. 205) and "can be represented in space in multiple ways" (ibid., p. 208). At the same time, these network subjectivities, conceived as nodes, remain partial, because the cooperation process only engages a partial function and perspective pertaining to them; here, again, the plurality of the "net/work subjectivity/ies" is in evidence:

> For (human) subjectivity in particular, this means that on the one hand it appears as a point of intersection and aggregation of multiple linkages, which, for example, can be largely condensed into one point in the database entry. If, however, we take this to its logical consequence, a concept of this kind implies the fragmentation of subjectivity into a multi-layered configuration of distinct relationships. (ibid., p. 256)

Accordingly, "subjectivity" in the digital condition means the interweaving of technical and non-technical actors to form "net/work subjectivities", which we might describe as "pluralised linkages" (ibid., p. 266). As such, it appears as a fragmented plural, reflecting multiple different loci, morphological forms, or perspectives, rather than as a singular unit. As "nodes", network subjectivities are variably integrated into the open and expandable reference structure of the network and are just as flexible as the relational "lines" that interlink them. They therefore effectively consist in intersections; we may conceive of them in analogy to "the basic principles of the internet service World Wide Web" (ibid., p. 265), which enables "the bringing together of dimensions without centring and standardisation" (ibid., p. 266).

Alongside this, Paulitz draws our attention to another aspect of network subjectivity in digital culture. The practice of network formation, understood as "constructive processes of networking" (ibid., p. 209), is another locus of overlap between technology and sociality/subjectivity. The interviewees conceive of the cooperation of partners in a project and the advancement of their qualifications as co-constructions which complete the technical construction of the network structure (cf. ibid., p. 258). These practices, producing the net/work as "technologies of networking", are "ultimately anchored in subjective self-relationships"; accordingly, Paulitz calls upon Foucault here in perceiving net/work subjectivities as "technologies of the social self" (ibid., pp. 20, 269). This framing of the findings from a governmental perspective disrupts her ethnographic researcher's point of view, but it simultaneously enhances the concept of network subjectivity by adding a processual component. From this point of view, the digital condition itself consists of incessant practices of networking, that is, the creation of networks.

We will again encounter this simultaneity of result and process in the definition of network subjectivity (cf. ibid., pp. 260, 268) in the two approaches I will go on to discuss, which likewise connect technical and non-technical actors. Tanja Paulitz's ethnographic study thus provides resonant points of reference for the capacity of the concept of 'network subjectivity' to help us approach digital culture. My reading of Paulitz' work is that it primarily draws attention to the extent to which digital technologies have become integral to everyday life, including to subjects' communicative self-concepts. One question remains open regarding Paulitz' research design. She professes her interest in avoiding a homogeneous representation of the various "net/work subjectivities". However, having examined two different projects, she ultimately pulls her findings together under one overarching concept of "net/work subjectivity/ies". It might have been more consistent, in the light of Paulitz' objective, to have distinguished specific conceptions of "net/work subjectivity/ies"; the use of the plural, however, goes some way towards resisting any tendency towards homogeneity.

Network Subjectivity as Affect-Based, Object-Mediated Collectivity or Agencement (Wiedemann)

While the governmentality framing, which Paulitz adds to her empirical analysis of network subjectivity in digital culture, serves rather as an outlook, the approach taken by the journalist Carolin Wiedemann in her doctoral thesis of 2016 draws explicitly on Michel Foucault, whose account of power is currently of considerably greater centrality in education studies; this referentiality to his work therefore invites us to engage more closely. Supplementing Foucault's philosophy with that of Gilles Deleuze, Wiedemann's analysis of the "Anonymous" collective, a decentralised international association of 'hacktivists' that rose to notable prominence in 2010 in connection with its "Operation Payback" to support Wikileaks, conceives of this grouping as an instance of affect-based, object-mediated, interactive collectivity.

Wiedemann, alongside her interest in Foucault's "search for new forms of subjectivity" (Wiedemann, 2016, p. 36), perceives Deleuze's concept of *agencement* (agency) as a neo-materialistic refinement of Foucault's power analysis based on interactivity (cf. ibid., p. 38).[13] Attempting to understand the interaction "of various variables involved in Anonymous and their collectivisation as the emergence of a common *agency*" (ibid., p. 40; italics in original), she terms this phenomenon/process "subjectivity" or "subject" (ibid., p. 184), which leads me to read her contributions as a response to Foucault's search for a new subjectivity that appears in this instance as a "collectivity" that does not fit conventional concepts of "collective identity" (ibid., pp. 50, 117, 147).[14] Read this way, Wiedemann's analysis of Anonymous, the hacker collective that is impossible to pin down, might contribute to our understanding both of a "new form of collectivity" (ibid., p. 41) and of a new type of 'network subjectivity' in digital culture.[15]

In the term "collectivity", the multiplicity inherent to "interspecies reciprocity" (Knorr Cetina, 2005, p. 589)—a technology-based, interactive component of this form of subjectivity that Wiedemann (2016, p. 201) calls "intraactions" (Barad)—is more explicit than in the plural "net/work subjectivities" of Paulitz' analysis. Paulitz and Wiedemann would agree, however, that technical and humanoid actors no longer merely interact, but, potentially at least, move together to form the singular of a new manifestation of subjectivity or, put differently, agency. The specific characteristic of Anonymous' form of collectivity lies in the unpredictable cooperation it embodies, "which has no purpose beyond the spontaneous experience of collectivity in the sense of a shared efficacy, beyond the pleasure of creating together—and which would not exist without the internet" (ibid., p. 14). In concluding that the Anonymous phenomenon remains as mobile as the ICT-generated network structures of social media communications themselves, Wiedemann, like Paulitz, suggests that the subjective and the technical-material sides of this collectivity's network-like form are essentially of the same stuff.

Again like Paulitz, Wiedemann pays close attention to the processuality of what takes place as Anonymous continuously constitutes itself. As discussed above, Knorr Cetina had located the impulsive moment of this incessant process in the Lacanian dynamic of lack and desire—and, as I will go on to explore, Torsten Meyer places it in the concept of the Borromean link that links the real, the imaginary and the symbolic. Paulitz, by contrast, sees this process as unfolding in a correspondence of societal and subject-related forces which she encapsulates in Foucault's term "technology of the self". Returning to Wiedemann, I note that she gives two interpretations of the processual dynamic of this collective agency as a "force". In the first of these, following Foucault, she refers to the subjectivating "force field" (ibid., p. 22) of power in the biopolitical control society, which is coupled with self-management. This governmentality-based framing exposes more explicitly the links between subjectivity and, first, more historical perspectives on changes in the conditions of production, alongside, second, the current discourse around the "economisation of the social" (ibid., p. 23). These references were fundamental to Castells' analysis and also gained relevance to the digital condition in Stalder's citation of the knowledge economy. The second interpretation of the processual dynamic of collective agency utilises Deleuze's moment of "affection, from which the transformational forces within an *agencement* emanate" (ibid., p. 38, italics added). The concept of *agencement* ("agency"; in German the term is *Gefüge*) denotes and encompasses the specific form of a collectivity acting together as a subjectivity, but not bound to a single human subject: "The concept of *agencement* enables the theorisation of subversion beyond processes of subjectivation and reference to forms of interaction beyond only interpersonal interactions, which are based on processes of non-linear reciprocal affecting and being-affected" (ibid., p. 38, italics added). Wiedemann concedes that such states of affectedness also arise through auto-affection (as proposed by Knorr Cetina). We nevertheless note, proceeding from a neo-materialist perspective (as in the actor-network theory advanced by Bruno Latour, and also in the approaches of Karen Barad and Gilles Deleuze), that this affectedness is detached from specific subjects and bound to forces that circulate as a responsive phenomenon between (material) bodies or materialities (cf. ibid., pp. 149, 218, 242). This linkage process of emerging *agencement/* agency sustained by the moment of affection appears as a crucial component of the transformation implicated in *agencement*, or as the swarm-forming "force of affects" (ibid., p. 111) unleashed by networking logics. Affects therefore institute the collectivisation of individual elements into a swarm (ibid., p. 116).

*Agencement/*agency likewise conceives of technical media via which networking takes place "as actors in the process of becoming different and collective" (ibid., pp. 58, 122). This view includes infrastructural "levels of solidification and crystallisation, for example of algorithmic codes" (ibid., p. 38) as parts of this process, part of the concrete expression, form and visibility of the collective agency of Anonymous, thus structuring "the interplay of cultural practices and technical infrastructure" (ibid., p. 123). I perceive the emergence here of an incipient, temporary collective network subjectivity. The 'birthplace' of the Anonymous phenomenon—the message board of the social network 4chan, its anonymous users, their conversations and/or memes, and the rhizome-like structured collectivity coming forth from their interactions—temporarily become a collectively acting "subject" and form (to speak with Massumi) a network via "reciprocal affects" (ibid., p. 189).

Wiedemann's analysis of Anonymous proceeds beyond the purely digital space, pointing to intertwinements between online and offline activities and swarm-like phenomena of assembly such as Occupy and flashmobs (cf. ibid., p. 147), planned digitally and carried out in the real world, then returned to the digital sphere via documentation, sharing and storage, there to influence further real-world actions—a continuous, reciprocal passing of action between the spheres. In this context, she goes as far as to refer to a "swarm network" (ibid., p. 162). The appeal of her account to the analysis of subjectivation processes in the digital condition/the informational network society is twofold: First, the metaphor of the network may help conceptualise technical media and human individuals as collective actor-subjects, as an *agencement*; second, Wiedemann's interconnection of online and offline processes may offer a description of network subjectivity as a collective structure.

I conclude this section by noting an unanswered question as to the "forces of affects" with their capacity to form collectivities. Wiedemann asserts that affirmations establish the cohesive 'power' of the logic of networking and thus form the *agencement*, the interaction of Anonymous as a technical/humanoid collective-subject. The affectivity of human individuals, however, is bound to a sphere of bodily presences defined by visual, acoustic, and tactile phenomena. It does not seem quite clear to me how mutual affections, thus detached from subjects, can enter into a sphere of interaction and communication mediated by technical systems and governed by completely different systems of signs, be they constituted of meaning or code. How can we present a coherent idea of an affection of *agencements* as an in between that does not ultimately re-disintegrate into the distinct entities of the acting human being and the executing technical infrastructure?

The Network sujet *as a Borromean Link (Meyer)*

Knorr Cetina's work demonstrates that the metaphor of the network may exercise notable appeal to psychoanalytical models of subjectivity. Torsten Meyer is another theorist who eventually engages with psychoanalytical paradigms in this context. He initially embarks on his recent analysis of the "network *sujet*" (French, "network subject") from a perspective informed by art and media theory and in explicit reference to the anticipated or effected changes in technologically mediated human interactions associated with the digital age. His account draws on the concept of "mediology" as proposed by the French philosopher and activist Régis Debray, which posits the far-reaching cultural and social impact of epochal media technologies such as language, writing or books and their respective semiotic systems. Debray calls the present age, with its predominance of digital signals emitted and received by computers, the "videosphere" or "hypersphere" (Meyer, 2011, p. 15). Meyer concurs with these thoughts, asserting that "[c]hanges in mediality lead to changes in subjectivity" (ibid., p. 39).

Meyer (2018, p. 39) notes the existence in the current age of a "net/work-shaped mediality", affirming the analogous nature of technical/material to subjective forms which Castells rejects: "Present-day mediality is characterised by networks—by real, material networks of devices and by virtual, metaphorical networks in our thoughts" (ibid.). In this respect, Meyer's first proposal is to abandon the Cartesian model of subjectivity that determines modernity, predicated on a "dualism of I and the world, subject and object" (ibid., p. 40); he instead calls on Lacan's model, crucial to Knorr Cetina, of the psychic apparatus that underlies the dynamics of lack and desire. This model consists of the three registers of the real, the symbolic and the imaginary, intertwined to form a "Borromean link" (literally "Borromean node", originally "Borromean rings"), in which Meyer perceives a representation of "network-shaped subjectivity" (ibid., p. 40). The specific form of this linkage (e.g. of rings or loops) epitomises its essential property—that is, that none of the parts can be removed without the other parts falling apart.

Building on this, Meyer proposes visualising the altered interaction structures of the hypersphere, which Knorr Cetina had termed "interspecies reciprocity", using an installation by the artist Tomás Saraceno consisting of connected steel cables that form a walk-in net modelled on spiders' webs. The intent of this structure is the mimesis of a "vibrant net/

work of relationships, resonances and synchronous communication" (ibid., p. 55) between individuals and (technical) objects. By this analogy, media-driven subjectivities, conceived as "Borromean links", would thus intertwine in the hypersphere to form a 'Borromean net/work'.

As attractive and inspiring as these visions may initially appear to the analysis of subjectivity in digital culture, caveats present themselves: the core of the Borromean link seems to me to lie in its non-nodeness—rather than a fastening, it comprises and effects an indissoluble entanglement without beginning or end. It stands for itself and, unlike a node, therefore provides no support for anything but this self-relation. To return to Lacan, the real, as one of the three subject-constitutive registers alongside the symbolic and the imaginary, is not available to the subject in positive form, but only shows itself as a withholding/withdrawal within the symbolic. The productive and processual dynamic of subjectivity at which Meyer thus arrives, and as also features in Paulitz' governmentality framing and in Wiedemann's Deleuzian references, constitutes itself, as in Knorr Cetina's analysis, in the negative concept of lack.

Even if a Borromean link were, in principle, infinitely extendable to a Borromean net via processes of linking,[16] the somewhat 'static' image implies the model's potential for use to describe both the dynamics of a structural description of our psychic set-up in this regard and the processual dynamics of media-driven interactions. Conversely, the image of the spider's web installation seems to deprive the Borromean link of its quintessential property in the present context: Its interwovenness is precisely not that of a fabric connected by nodes (or web-like adhesions). The original Borromean rings, also representable by loops, can be moved around within their linked state, but no one ring can be removed without the entire linkage dissolving. Applied to the dynamics of subjectivity, this would mean that a subjectivity defined by the Borromean entanglement remains mobile due precisely to the negativity or deficiency caused by the real that withdraws/withholds itself within the symbolic (and ultimately cannot be brought to rest even in imaginary wish fulfilments). As a metaphor, the Borromean link seems able to transport the dynamics of an 'intraactive' subjectivity—but does this likewise hold for a net/work-shaped mediality? My sense is that this characteristic may potentially be lost in the visualisation of the network metaphor as a spider's web.[17]

Meyer's location of subjectivity in the hypersphere, proceeding from a media studies standpoint, corresponds with Knorr Cetina's sociological description of postsocial subjectivity insofar as both decisively include the relationships between people and things—which are also key to the accounts given by Paulitz and Wiedemann. This is also the reason why Meyer now abandons the "subject" for the *sujet*, the "'theme', 'material', 'motif', etc." (Meyer, 2018, p. 40). The "tools of the symbolic, for example search engines, advertising algorithms, book recommendations, dating sites etc." (ibid., p. 56) then become part of an extended network subjectivity, without the possibility of imputing to these manifestations of artificial intelligence a formation of meaning or intentional action bound to human individuality. In this way, alterity, i.e. the "radical strangeness of the *imagination* of such hyper-complex computer systems" (ibid., p. 57; italics in original), is reserved not only for human subjects and relationships, but also for things and object relations in the hypersphere. This view of the connection between digitality, mediality and alterity raises crucial and complex issues for education studies.[18]

4 Conclusion

This chapter proceeded from the assumption that approaches to describing present-day subjectivity correspond to specific frameworks from social theory. Accordingly, it examined the intersections between discrete interpretations of contemporary society, drawing on the disciplines of sociology and cultural studies, which define our current condition as "postsociality" (Knorr Cetina) or as "digital culture" (Stalder). This analysis suggests that the concept of the "network" has the capacity to combine approaches from social theory and the theory of subjectivity; in analogy to the "network society", I therefore proposed the metaphor of 'network subjectivity' and generated three distinct descriptions of the concept in engagement with three studies from different areas of research.

Each of these studies detach subjectivity from its classic adherence to the single human individual. The ubiquity of informationalisation as noted by Castells urges us to take seriously the constitutive relationship between human, non-human and technical actors and to ascribe agency and interactivity to these "net/work subjectivity/ies" (Paulitz), "collectives" or *agencements* (Wiedemann), or the "network *sujet*" (Meyer). In all three studies, a paradox emerges when the concepts developed in the metaphor of 'network subjectivity' appear as results or ends, yet are simultaneously described as processual. I note here a parallel to the "paradox of subjectification", which Bröckling (2016, p. 1) analyses from the perspective of governmentality studies as a "contradiction between self-constitution and antecedent constitution".

The affective moment takes on differing relevance in each study, attaining no independent role in Paulitz's analysis, but occurring in Wiedemann's work as a constitutive force of networking and in Meyer's study as at least implied in the psychoanalytical subject theory on which the author draws. Ultimately, in all three concepts, a subjectivity conceived in this way becomes a plural singular that stands for a collective structure. It remains constitutively tied to the volatility of the affects and/or interactivities that produce it, to their processuality and productive dynamics, incapable of progressing to an identifiable, representative entity. Thus, for example, it can present itself in, and as acting in, virtual workspaces (Paulitz), appear as a swarm (Wiedemann) or in interactivity with/of responsive media (Meyer). Each of these manifestations extend beyond the online world and occur as offline phenomena.

None of the three approaches I have analysed here under the metaphorical umbrella of 'network subjectivity' can stand alone as conclusively elaborated theories of subjectivity in the digital condition. They do, however, share a striving to make accessible the changes that information technologies are bringing to bear on our practices of subjectivation and our conceptions of ourselves in our age. In so doing, they resist the idea, currently dominant in media literacy education and media pedagogy, that the competent, media-literate, autonomously acting computer user is in control of that computer and of internet processes. Indeed, they instead indicate that we have always been part of the technologies we use and that the influence of this use on our thoughts and actions is equal to that of the reverse. This insight tallies with many observable phenomena in technology use, in the fields, for example, of gaming and cosplay, and with the evidently libidinously charged fascination exerted by digital technologies.

This chapter does not intend to claim sole or indeed even dominant status for 'network subjectivity' as a concept of subjectivation in our time; if it were to do so, it would not take long for the observation to surface that the technical and social components of this metaphor find themselves often enough in opposition to one another. However, the attempt I present here to identify an appropriate concept of subjectivity may shed light on specific, frequently observed interactive media phenomena in our digitalised day-to-day, examples being experiences with communications in social media or interactions with devices such as smartphones or wearables. As Tanja Paulitz has noted, while the incorporation of such phenomena into a discourse on the opportunities and risks of media use may seem apposite, it misses out the aspect of subjectivity. This chapter represents an experiment in placing this aspect at the heart (or hearts) of the discourse.

Notes

1. I have chosen to use the terms 'digital condition' or 'digital culture' in this chapter rather than 'post-digital', although the latter might appear more in line with the term 'postsocial' (Knorr Cetina) that I discuss in the section that follows. 'Post-digital' is similar in emphasis to the 'digital condition' as defined by Stalder (2018, p. 9); both terms argue that digital media and technologies have migrated deep into material infrastructures, everyday life and social interactions. In contrast to 'digital condition' or 'digital culture', however, the term 'post-digital' remains, in its very structure, fixated on these technologies and carries the inevitable connotation that something has been 'left behind' (see also Jörissen, 2018, who uses 'post-digital' in a discussion of subjectivity from a perspective informed by cultural and media studies).
2. In 2011, Rob Wilkie had published a book also entitled *The Digital Condition*, which is not cited by Stalder. Providing a critique of the digital condition that draws on the theory of class, this classic Marxist approach is contrary to Stalder's cultural studies-influenced perspective. Stalder's book was first published in German in 2016 with a title that, rendered literally, would be "Culture of Digitality". This chapter uses the terms 'digital condition' and 'digital culture' synonymously.
3. On the classification of Castells as a cultural theorist, cf. Nollmann (2011).
4. This is also evident in the interview for *First Monday* (https://firstmonday.org/ojs/index.php/fm/article/view/9409/7574, last accessed 09.08.2020).
5. This is also the view held by Jörissen (2016, p. 232), who therefore suggests the use of the network concept in the light of a theory of difference.
6. We may also interrelate Lacan's structural description of a constant effort to maintain libidinal object relationships which incorporate power and economic interests with analysis of the consumer industry. Cf. Schenk and Hoffarth (2018) on Slavoj Žižek's reformulation of a critique of ideology using these Lacanian figures. Mayer and Schenk (2018) explore Žižek's adaptation of Lacan in greater depth.
7. Bruno Latour's term "interobjectivity" (Reckwitz, 2008, p. 166) has probably been used more widely.

8. This could also be the reason why the study of subjectivity does not amount to an established independent field of network analysis or research (cf. von Kardorff, 2019, p. 108).
9. On the relevance of network research in education studies, a discipline not focused on subjectivity and tending towards a largely instrumental view, cf. Berkemeyer and Bos (2010).
10. In the interest of education and with reference to a changing modern sociality conceived of in terms of a theory of difference, Jörissen (2016, p. 232) likewise supports this approach. I pursue this aspect elsewhere (Schenk, forthcoming).
11. The original quotations contain a number of italicised passages, which hinder legibility in the present context and are therefore omitted.
12. These are key concepts in quantitative network research. "Nodes" represent the actors in the network, "lines" their relationships (or lack thereof) to one another. These concepts enable the mathematical and graphical representation of the relationship structures governing social networks (see Brandes, 2010).
13. The shift from Foucault to Deleuze also relates to an impression I gained while reading that relates to the systematic fit of the concepts and perspectives used, and especially the analytical fuzziness of the Deleuzian concepts. The thesis was a cumulative collection of publications rather than a monograph, and so each piece reveals a different emphasis. I also perceive a shift in the author's interests during the doctoral process, which neither the detailed introduction nor the final discussion can entirely bring under one conceptual umbrella.
14. On the issue of empirical research on collectives as identities—without reference to Wiedemann—see Schenk (2019).
15. This linkage of subjectivation and collectivation brings together sociological and educational perspectives in contemporary debates (cf. Alkemeyer et al., 2018).
16. Illustrations of different entanglements appear in The Knot Atlas (http://katlas.math.toronto.edu/wiki/L6a4, last accessed 09.08.2020).
17. In an earlier publication, Torsten Meyer (2015, p. 94) had proposed conceiving of the subject as a "*sujet*" in "educational practices based on collaborative and networked socio-technical processes in global, digital communication networks" (Meyer, 2015). This would be the starting point for the formation of a network subjectivity as a "*sujet*" in relation to "the network itself and the communities that *form within* it" (ibid., p. 114).

18. Knorr Cetina's (1997, p. 16) analysis of how the biologist McClintock describes her relationship to the organisms under study, or how objects themselves appear as processual and integrated into the dynamics of lack and desire (cf. Knorr Cetina, 2005, p. 589), also leads to this insight of alterity. Accordingly, both Knorr Cetina (2005, p. 590) and Meyer (2018, p. 61)—likewise Wiedemann (2016, p. 62)—link their observations, rather by the by, with post-humanism, currently just as central a concept to education studies as is the associated idea of alterity as an imperative of justice (cf. Wimmer, 2019).

References

Alkemeyer, T., Bröckling, U., & Peter, T. (Eds.). (2018). *Jenseits der Person. Zur Subjektivierung von Kollektiven*. Transcript.

Berkemeyer, N., & Bos, W. (2010). Netzwerke als Gegenstand erziehungswissenschaftlicher Forschung. In C. Stegbauer & R. Häußling (Eds.), *Handbuch Netzwerkforschung* (pp. 755–770). Springer VS.

Bröckling, U. (2016). *The entrepreneurial self. Fabricating a new type of subject*. Sage.

Brandes, U. (2010). Graphentheorie. In C. Stegbauer & R. Häußling (Eds.), *Handbuch Netzwerkforschung* (pp. 345–353). Springer VS.

Castells, M. (2010). *The rise of the network society. The information age* (Vol. 1, 2nd ed.). Wiley-Blackwell.

Cramer, K., Fulda, H. F., Horstmann, R.-P., & Pothast, U. (Eds.). (1990). *Theorie der Subjektivität*. Suhrkamp.

First Monday. (2018, August 28). *The (Post-)digital condition – An interview with Felix Stalder*. Retrieved February 26, 2020, from https://firstmonday.org/ojs/index.php/fm/article/view/9409/7574

Hepp, A. (2010). Netzwerk und Kultur. In C. Stegbauer & R. Häußling (Eds.), *Handbuch Netzwerkforschung* (pp. 227–235). Springer VS.

Holzer, B. (2006). *Netzwerke*. Transcript.

Jörissen, B. (2016). Zur bildungstheoretischen Relevanz netzwerktheoretischer Diskurse. In D. Verständig, J. Holze, & R. Biermann (Eds.), *Von der Bildung zur Medienbildung* (pp. 231–255). Springer VS.

Jörissen, B. (2018). Subjektivation und ästhetische Bildung in der post-digitalen Kultur. *Vierteljahrsschrift für wissenschaftliche Pädagogik, 94*(1), 51–70.

Knorr Cetina, K. (1997). Sociality with objects. Social relations in postsocial knowledge societies. *Theory, Culture and Society, 14*(4), 1–30.

Knorr Cetina, K. (2005). Postsocial. In G. Ritzer (Ed.), *Encyclopedia of social theory* (Vol. 2, pp. 586–591). Sage.

Mayer, R., & Schenk, S. (2018). Žižek. In P. Smeyers (Ed.), *International Handbook of Philosophy of Education* (Vol. 1, pp. 457–469). Cham.

Meyer, T. (2011). Medien & Bildung. Institutionelle Kontexte und kultureller Wandel. Einleitung. In T. Meyer et al. (Eds.), *Medien & Bildung. Institutionelle Kontexte und kultureller Wandel* (pp. 13–25). Springer VS.

Meyer, T. (2015). Ein neues Sujet. In B. Jörissen & T. Meyer (Eds.), *Subjekt Medium Bildung* (pp. 93–116). Springer VS.

Meyer, T. (2018). Das Netzwerk-Sujet. In F. Kiefer & J. Holze (Eds.), *Netzwerk als neues Paradigma? Interdisziplinäre Zugänge zu Netzwerktheorien* (pp. 39–63). Springer VS.

Nollmann, G. (2011). Manuel Castells: Kultur, Technologie und Informationsgesellschaft. In S. Moebius &·D. Quadflieg (Eds.), *Kultur. Theorien der Gegenwart* (2nd ed., pp. 635–644). Springer VS.

Paulitz, T. (2005). *Netzsubjektivität/en. Konstruktionen von Vernetzung als Technologien des sozialen Selbst.* Westfälisches Dampfboot.

Reckwitz, A. (2008). *Subjekt.* Transcript.

Ricken, N., Casale, R., & Thompson, C. (Eds.). (2019). *Subjektivierung. Erziehungswissenschaftliche Perspektiven.* Beltz Juventa.

Schenk, S. (2019). Das 'Wir' der Proteste. Zum Verhältnis von Identität und Differenz in einer poststrukturalistischen Empirie. In A. Schäfer & C. Thompson (Eds.), *Gemeinschaft* (pp. 149–171). Schöningh.

Schenk, S. (forthcoming). Experimentieren in der Netzwerkgesellschaft. Praktiken der Bildung in der Kultur der Digitalität. In K. Zehbe & M. Zulaica (Eds.), *Rhetoriken des Digitalen.* Springer VS.

Schenk, S., & Hoffarth, B. (2018). Konsum und Kritik. Ideologiekritik als Performance inszenierten Genießens bei Slavoj Žižek. In C. Bünger, O. Sanders, & S. Schenk (Eds.), *Bildung und Politik nach dem Spätkapitalismus* (pp. 155–182). Argument.

Stalder, F. (2008, May 18). *About.* Retrieved April 27, 2020, from http://felix.openflows.com/node/4

Stalder, F. (2018). *The digital condition.* Polity.

The Knot Altas. (2009, 27 May). *L6a4.* Retrieved March 7, 2020, from http://katlas.math.toronto.edu/wiki/L6a4

von Kardorff, E. (2019). Castells, Manuel (1996, 1997, 1998): The information age: Economy, society and culture. In B. Holzer & C. Stegbauer (Eds.), *Schlüsselwerke der Netzwerkforschung* (pp. 105–109). Springer VS.

Wiedemann, C. (2016). *Kritische Kollektivität im Netz. Anonymous, Facebook und die Kraft der Affizierung in der Kontrollgesellschaft.* Transcript.

Wilkie, R. (2011). *The Digital Condition. Class and Culture in the Information Network*. Fordham University Press.
Wimmer, M. (2019). *Posthumanistische Pädagogik. Unterwegs zu einer poststrukturalistischen Erziehungswissenschaft*. Schöningh.
Winter, R. (2010). *Widerstand im Netz. Zur Herausbildung einer transnationalen Öffentlichkeit durch netzbasierte Kommunikation*. Transcript.

CHAPTER 3

School of Data and Shifting Forms of Political Subjectivity

Valentin Dander and Felicitas Macgilchrist

Abstract Digital media are increasingly 'data media' and data media are involved in various forms of political activism. This chapter reconstructs political subjectivities around figurations of the 'digital citizen' within the field of (open) data activism. The authors draw on interviews, document analysis and concepts from modern and post-sovereign political theories of subjectivation to explore the transformative educational work of the *Datenschule* (School of Data) project, focusing on the intersection between open data and anti-discriminatory activism. The chapter suggests that although School of Data explicitly positions its work as supporting 'skills' acquisition (data literacy), indicating a modernist understanding of subjectivity, the project also generates an understanding of political

V. Dander (✉)
Fachhochschule Clara Hoffbauer Potsdam, Potsdam, Germany
e-mail: v.dander@fhchp.de

F. Macgilchrist
Georg Eckert Institute for International Textbook Research, Braunschweig, Germany

Institute of Educational Science, University of Göttingen, Göttingen, Germany
e-mail: macgilchrist@leibniz-gei.de

P. Bettinger (ed.), *Educational Perspectives on Mediality and Subjectivation*, Palgrave Studies in Educational Media,
https://doi.org/10.1007/978-3-030-84343-4_3

subjectivation as a multiplicity of distributed transformative processes, entangling data literacy with power structures, data-related and organisational practices.

Keywords Datafication • Open data • Political subjectivity • Data literacy

1 Introduction

Through the digital data traces we leave every day, we have become 'transparent subjects'. If 'we are data', as the title of a recent book suggests (Cheney-Lippold, 2017), then algorithms determine our identity, our friendships and our access to news, information and ads; algorithms assign us to race, gender, and class categories, and regulate which forms of citizenship and politics we can enact. In this view, the central political effect of data is to render us powerless in the face of the corporations that extract our behavioural data in pursuit of profit and/or the governments which profile and police us at ever more granular levels by using automated data tools (Couldry & Mejias, 2019; Zuboff, 2019). At the same time, the ubiquity of digital data enables other political effects, such as their use to enhance equality or the centring of data justice and data activism in collective efforts for change.[1]

Rooted in a post-foundationalist and generative orientation to (critical) discourse analysis, this chapter explores the political subjectivities which are—or can be—associated with educational activist projects on data in public life. It does so by first, in Sect. 2, reflecting on cultural and political theories of political subjectivity/subjectivation which offer important insights into digital citizenship and data activism in today's datafied societies. We then turn in Sects. 3, 4, and 5 to a 'worked example', analysing the discourse of non-profit data analytics projects, run by School of Data Germany (*Datenschule*), which teach data literacy for anti-discriminatory practices and other socially progressive purposes. We tease out the priorities invoked in the School of Data's website, interviews with a member of the project's team, and selected publications on open data, data activism and anti-discrimination work. To conclude, Sect. 6 returns to the theories of political subjectivities presented in Sect. 2, and reflects on the broader implications of these understandings for data justice work.

Overall, the chapter seeks to identify which conceptualisations of 'becoming a political subject' are at work here. While on one level, the materials issued by School of Data Germany explicitly position this process as skills acquisition (data literacy), a broader, less controllable understanding of political subjectivation as relational self-transformation also emerges. We thus conclude by suggesting that although the contemporary discourse of 'data literacy' often foregrounds limited, individual 'skills', set apart from their context, the ways in which specific projects enact data literacy demonstrate more far-reaching, power-related, collective, organisational, relational and distributed data practices. Our analysis suggests the need to address these data practices more explicitly when theorising data literacy and/or when developing educational data projects.

2 Political Subjectivities, Digital Citizenship and Data Activism

Current scholarship on data activism assumes, we suggest, various figurations of the '(digital) citizen'. Drawing on conceptualisations of political subjectivity in recent cultural and political theory, this section teases out three approaches to digital citizenship.

Individual Political Subjectivity: 'I am a citizen, I am a political subject!'

Classic liberal conceptions of political subjectivity in (late) modernity posit the citizen as a sovereign, rational, autonomous and self-determined individual (Mack & Gaus, 2004, p. 116; Rawls, 1996, p. 306). One or more nation states confer 'citizenship' on individuals, endowing them with civil, political and social rights. This 'thin' conception of citizenship, as Saward (2006, p. 403) puts it in his entry in *The Oxford Handbook of Political Theory*, circulates around one 'key specific space' (Saward, 2006, p. 403), the polling booth, which largely discounts activism as part of citizenship. Approaches including deliberative democratic theory have challenged this conception, extending it to forums of "talk, dialogue, reasoning together, becoming informed together, and making decisions" (Saward, 2006, p. 404). This view reads citizenship as not simply 'given' (at birth, or later through formal legal applications), but also 'enacted', and resonates in an understanding of 'the virtue of active citizen participation in community affairs' (Saward), and in the idea of democratic discourse through identifiable, "public contestation" (Schaal & Heidenreich, 2006, p. 122). This enacting is key to understanding political subjectivity in democracies, as citizens exhibit political agency and seek to gain an enlightened understanding of political issues, and act as individuals to exercise control over their polity through political representation (Isin & Ruppert, 2015).

Current discussions of *digital* citizenship frequently adopt a similar approach to political subjectivity, with 'digital citizens' seen as sovereign, relatively autonomous individuals who achieve citizenship through their practices, including "using technology to make [their] community better; [e]ngaging respectfully online with people who have different beliefs [...] [u]sing technology to make [their] voice heard [...] and to shape public policy", and "[d]etermining the validity of online sources of information" (ISTE, 2019). This digital citizen has the 'skills' to "find, evaluate, and share information responsibly, engage in constructive conversation with others from diverse backgrounds, and [...] ensure their online participation is safe, ethical, and legal" (Gleason & von Gillern, 2018, p. 200). In this context, "[d]igital citizenship can be described as the norms of appropriate, responsible behavior with regard to technology use" (Ribble & Bailey, 2007, p. 10). This perspective conceives of political subjects as rational individuals who enact their digital citizenship, and thus their political agency, by behaving responsibly, inclusively and ethically in digital spaces.

Similarly, some recent approaches to data activism emphasise the possibilities for individual political agency that emerge through data assemblages; they include the promotion of a critical consciousness, grassroots data literacy and a critical imagination for creating alternative ways of engaging with data that fall outwith their elite or capitalist exploitation (Gutiérrez & Milan, 2017). This work, often drawing on phenomenology, aims to engage with individuals' perceptions, experiences and reflections of the datafied world (Couldry & Powell, 2014; Kennedy, 2018; Sander, 2020). While vitally important for identifying how datafication is not only enforced top-down on technology users, but is made and unmade in everyday data-based interactions, this research also suggests—sometimes explicitly, sometimes implicitly—an understanding of the political subject as a relatively autonomous, reflective, deliberative, intentional individual, operating within representative democracy. Although this approach reflects on communities, alliances and collaboration as part of data activism, its analytical interest lies in the sense-making, critical orientation and actions of the *individual* living with data.

Collective Political Subjectivity: 'We are here, we are loud, we take up space as a political subject!'

Alongside this subject-centred emphasis, an alternative understanding of data activism emerges in other approaches that explicitly conceptualise the political subject as a contingent, relational performativity that comes into being through the collective presence of voices and bodies in public spaces. In illuminating this conception, we draw on post-foundational writing that sees the subject as emerging from (constituted, but not limited by) situated, embodied, normative sociality.

Two key thinkers in this regard are Jacques Rancière (2001, 2013) and Judith Butler (1993, 2015). Their approaches offer us two central ideas for understanding data activism today. The first of these is the 'distribution of the sensible', which describes the "system of self-evident facts of sense perception" that enables something to be seen and heard, thought or said, made or transformed (Rancière, 2013, p. 7). The distribution of the sensible thus enables or forecloses perception, communication and action, consequently shaping the potential of (data) activism as framed in democracy: "Politics revolves around what is seen and what can be said about it, around who has the ability to see and the talent to speak, around the properties of spaces and the possibilities of time" (Rancière, 2013, p. 8). In this view, the political participation offered to the demos within representative democracy appears contained, domesticated, tamed; its primary function is to include and exclude, to 'police', in the sense of keeping people in their place (Rancière, 1999, p. 28f.). A core objective of data activism would thus be to 'redistribute' the sensible, i.e. to change what and who is visible or invisible, audible or inaudible, and hence who has a 'part' in politics (Rancière, 1999, p. 9).

A second central idea is the role of bodies in activism, and how data activism relates not only to voice but also to physical presence. Butler refers to "*Bodies* in Alliance", *doing* street politics 'between' themselves (Butler, 2015, p. 77). In this understanding, activist practices primarily operate on the level of the "performativity of the body" (Butler, 2015, p. 83). Contesting suggestions that protest has migrated online, Butler writes: "Although some may wager that the exercise of rights now takes place quite at the expense of bodies on the street, that twitter [sic] and other virtual technologies have led to a disembodiment of the public sphere, I disagree" (2011, n.p.).

Interestingly, there is a slight shift in Butler's take on mediality between this 2011 lecture, 'Bodies in Alliance and the Politics of the Street', and the publication of the book which grew from the lecture. In the book, the sentence quoted above on 'twitter and other virtual technologies' ends with "I would disagree in part" (Butler, 2015, p. 94). The claim that bodies anchor political events remains, but this shift, as well as other passages, are indicative of the fundamental entanglement between bodies in physical presence and bodies in media. In this reading, then, the political event comes into being, not only between the bodies, but even more so *between* the bodies in alliance *and* their mediation (Butler, 2015, p. 92, 94).

Again, the implication for data activism is a focus on increasing the visibility and audibility of those bodies and voices usually excluded from the political (see, for example, Daly et al., 2019; Dencik et al., 2019; Milan & Treré, 2019). When, for instance, the 'digital citizen' enacts itself online by witnessing, hacking, and commoning, it uses the 'performative force' of legal, written speech acts, such as human rights declarations or national constitutions, to 'become' a citizen (Isin & Ruppert, 2015, p. 122f.). The 'figure of a citizen as yet to come' thus has a framework for the right to claim rights (Isin and Ruppert, p. 123f.). Or when, for example, lower-income families, traditionally excluded from being seen or heard as political actors, use data to raise their voices about inequality, or to show the extent to which schools discipline African American students more harshly than white students, then data undergirds the performative action of raising allied voices and making bodies present in public spaces (Macgilchrist, 2019, p. 83).

Obfuscational Political Subjectivity: 'I am invisible and anonymous, but I still have political agency!'

One way of radically rethinking subjectivity in the context of the digital is by questioning the apparently self-evident link between political subjectivity and the process of being seen or heard. One line of argument, for instance, defines the emergence of a political subject today through enacting a "right to opacity" (Birchall, 2016) and learning to "obfuscate" one's data traces (Brunton & Nissenbaum, 2015). Anonymity, in this sense, is not a negative practice of hiding or protecting oneself from censure; it is a positive political practice of acting anonymously *with the express purpose* of creating new forms of political subjectivity. This type of political subject destabilises traditional conceptions of political participation and democracy, in which the subject performs its visibility, raises its voice, and engages in open, transparent public debate (de Lagasnerie, 2017, p. 57f.).

Some anonymous online practices, such as distributed denial-of-service (DDoS) attacks, translate the physical sit-in protest into the online sphere. WikiLeaks and Anonymous are examples of how this political subject, acting anonymously, interrupts the dominant political order, which rests on the assumption that democracy is transparent, fair and participatory, yet which reveals itself as deeply implicated in reproducing socio-economic inequalities between those who participate in decision-making and those who do not. Anonymity, in this sense, is a proactive political practice of creating invisibility and, in so doing, of making the invisible visible. Invisibility is not 'hiding', it is acting purposefully. De Lagasnerie (2017) observes of WikiLeaks and Anonymous:

> By disconnecting the question of politics from the question of the public sphere, anonymity gives rise to a scene on which what one might call non-relational politics occurs: politics that is affirmative and radically emancipated from all ethical considerations—in other words, perhaps, pure politics. (de Lagasnerie, 2017, p. 75)

De Lagasnerie's argument is directed against observers who claim that anonymity is a cloak for criminality, rather than a legitimate form of political participation. In his view, anonymity democratises politics by enabling contributions to the flow of knowledge which fall outwith the prohibitions and censorship that limit the enacting of democracy in democratic states today (de Lagasnerie, 2017, p. 65). It enables subjects to speak from the inside and the outside at the same time (as do whistleblowers) and reduces the personal cost of political participation, such as prosecution and job loss. In this way, WikiLeaks, for instance, enables a '*split* subjectification' by providing the sociotechnical tools to simultaneously act within and against an institution (de Lagasnerie, 2017, p. 103; emphasis in original). Anonymity is, in this sense, a technique for casting off straightforward subjectivation (*désasujettissement* or 'desubjugation' in de Lagasnerie, 2017, p. 72), for disidentifying and disentangling oneself from the institution in which one works, for 'escaping citizenship' (de Lagasnerie, 2017, p. 99) as illustrated by 'Citizen Four', Edward Snowden.

Active creation of anonymity and invisibility takes place at the grassroots level through, for instance, various encryption mechanisms. These processes have been described as 'reactive data activism', i.e. an activism that reacts to the threats posed by data surveillance and mass capture of data (Milan & van der Velden, 2016). Even if de Lagasnerie does not refer to data activist practices, these may also count for him as "practices that are freer and more selective—more and more emancipated from the psychic hold of external and arbitrary constraints" (2017, p. 72). This notion of a freer practice, however, brings de Lagasnerie back to an understanding of political ('nonrelational') subjectivity which is strongly intentional and relatively autonomous. His account of anonymity and obfuscation in politics, while aiming to describe a novel political subjectivity, therefore returns us to a notion of ***individualised*** political subjects. This said, it does raise the question of the workings of digital data in other, perhaps more collective and contingent, forms of political subjectivity.

3 Worked Example: School of Data

To consider these theories of political subjectivity in concrete settings, we turn now to an extended example of a data activist project in Germany. After presenting our focal case in this section, we draw in the next two sections on a selective and intensive discourse analytical reading of website materials, two interviews and further publications. In doing so, we focus explicitly on *how* the texts account for their political and educational goals and how they relate to political subjectivities (see Kessl, 2005).

'School of Data' Germany (*Datenschule*)[2] was initiated in 2015 as one of several specific initiatives focusing on 'open data' run by the German chapter of the Open Knowledge Foundation (OKFde).[3] OKF is a non-profit organisation originally founded in 2004 in the UK, which champions the use of open data as material for empowering practices. The concept of open data, encompassing the subset 'open government data' (OGD), has gained substantial attention since the end of the 2000s and is closely related to the free software movement and other open-source and open access initiatives such as open culture and open educational resources (OER). Hintz et al. (2018, p. 132f.; see also Gutiérrez, 2018, p. 49ff.) classify open data campaigns as a form of 'proactive' data activism. The OKFde website outlines the principal characteristics of open data, in line with a generally accepted brief definition, as follows:

- Availability and access: Data should be available as a whole, at a cost no higher than the cost of reproduction, preferably as a free download on the Internet. The work should also be available in an appropriate and modifiable form.

- Use and re-use: The data must be made available under conditions that allow use, re-use and association with other data sets. The data must be machine-readable.
- Universal participation: Everyone must be able to use, re-use and re-distribute the data. There must be no discrimination against any persons or groups. The subsequent use may not be limited to individual areas (e.g. only for educational purposes), nor may certain types of use (e.g. for commercial purposes) be excluded. (Open Knowledge Foundation Germany, 2019)[4]

To a large extent, open data stems from public sources, as governments and administrations collect and work with information that is free from ownership and is considered highly relevant to the public. Key reasons given in support of opening up data are (1) transparency, (2) social and commercial value and (3) societal participation and engagement. The OKFde website explains:

> In a well-functioning, democratic society, citizens need to know what their government is doing. [...] Much of the time citizens are only able to engage with their own government sporadically [...]. By opening up data, citizens can stay better informed and be more directly involved in decision-making. This is [...] about **making a full 'read/write' society, in which citizens [...] are able to contribute [...]**.[5]

Authorities' objections to opening up data include the risk of making official secrets or personal/identifiable data widely available. Any disclosure of data, as in, for example, responses to Freedom of Information Act requests, will need to proceed after consideration of these concerns (Charalabidis et al., 2018, p. 1ff.). Scholars have criticised open data for opening the door to new public management methods in governance and to commercial interests that monetise open data (Dander, 2014, p. 45ff.), for depoliticising the public be turning them (us) into auditors (Birchall, 2016), or for operating as governing technologies by, for instance, not only publishing data but producing data through their 'aesthetic practices' (Ratner & Ruppert, 2019).

Other research about open data suggests that "this type of transparency has the potential to support the agency of datafied publics" and that it is taken as a means through which "the people should be considered again as the sovereign"' (Baack, 2015, p. 4f.; Baack, 2018, p. 169f.). However, even the members of the OKFde team have described this ideal of direct involvement between citizens and governments as overly simplistic:

> Even though the idea behind the democratization of information is to potentially allow everybody to interpret raw data, activists are well aware that the average citizen does not have the time and expert knowledge to do so. They recognize that their vision of empowerment through open data can only be realized with intermediaries that make raw data accessible to the public. (Baack, 2015, p. 6)

Given this lack of 'expert knowledge' on working with data, many have noted the need for educational projects in this area (cf. Wylie et al., 2019, p. 177f.). School of Data responds to this need by conducting workshops on data analytics skills for NGOs, young people and youth workers and by bringing data experts together with political activists to work on joint projects. At the time of our data generation, a team of five, consisting of one project manager/developer, one communications/press officer, two IT experts and one facilitator/coach, ran School of Data; as of June 2020, the website lists four permanent team members[6]). We conducted interviews[7] (see below) with the facilitator/coach who had conceptualised and predominantly run the workshops. He has a background in European Studies and Public Policy, but also some programming skills (I01:7:27–31). Although the team's roles seem to be clearly defined, all members work on multiple tasks, are politically involved and work closely with one another.[8]

The team ran and runs several projects, including 'Every School' (*Jede Schule*), an online database with open data on schools in Germany, 'Democracy Labs' (*Demokratielabore*), a joint project for media and civic education for young people and youth workers, and a collaboration with the environmental organisation 'Robin Wood' that focuses on environmental data.[9] Of particular relevance to this chapter is a joint project with ufuq.de, in which the two organisations co-developed a report on 'Using open data in anti-discrimination work' (see Sect. 5).

School of Data does not cooperate with businesses or other commercial actors; its focus is explicitly on non-profit organisations. The website notes: "We at School of Data Germany want to help realise data-driven projects: We cooperate with non-profit organisations based in Germany that campaign for positive social change."[10] Neither the organisation's mission statement nor any other part of its website specify a definition of *'positive'* in this context; our interviewee supplied a partial clarification by stating that cooperation with right-wing actors is not an option (I01:12:20–23). Further, irrespective of political leanings, School of Data does not accept requests to cooperate with any of the foundations formally aligned with political parties in Germany (be they conservative, socialist, green, etc.). Overall, then, the organisation's emphasis is on independent work with civil society actors on the progressive spectrum of political activity, within which context School of Data conducts numerous workshops each year (I01:18:17).

4 Modelling Data Literacy: From Skills to Empowerment

In the early years of open data, demands and discussions relating to technical, legal and political aspects of the issue dominated the discourse (Dander, 2014). We now note a closer focus on the specific educational and social practices of activists and various groups in civil society which seek to make open data actionable for counter-hegemonic positions. A long-term goal in this context is to make data experts/intermediaries unnecessary. It is in this context that we can read School of Data's principal societal and educational objectives, as listed on its website as 'knowledge transfer', 'social change' and 'community building'. The website contextualises the project's activities as follows:

> [N]on-profit organisations [...] often *lack* the required *skills* to use and turn data into valuable information. [...] In our workshops we *teach* the *skills* needed to process data and use related technologies in an *independent* and *self-determined* way. (Datenschule, 2021; emphasis added)

A similar description appears on the international School of Data website:

> We are a network of **individuals** and **organizations** working on empowering civil society organizations, journalists and citizens with **skills** they need to use data effectively [...]. We are School of Data and we believe that **evidence is power**. (School of Data, 2021; emphasis in original)

The descriptions refer to 'skills' as something people can 'lack' and that can be 'taught' in workshops, and postulate that one set of actors can be 'empowered' by another. With better skills, the texts suggest, people will be able to process data and use technologies independently and effectively. Alongside this understanding of skills as something that can be adopted by individuals in decontextualised settings and then implemented in participants' own projects, and somewhat contradictory to it, sits an alternative focus on 'organisations', community-building, campaigning and power. This foregrounds collaborative or structured work in groups or teams, within organisational and hierarchical societal structures.

Conceptually, School of Data uses two basic models developed 'on the job' in cooperation with academic researchers: the 'data pipeline' as a process model for data projects (I01:13:27ff.) and 'data literacy maturity' as a descriptive model for data literacy skills within organisations.[11] The data pipeline features prominently within the booklet *Jugendarbeit im digitalen Wandel* ('Youth work in the digital transformation'; Hahn et al., 2017, p. 39ff.). The model, considered a 'work in progress' being developed by the international School of Data community, currently consists of seven steps for literate engagement with (open) data, involving imperatives to define, find, get, verify, clean, analyse, and present data (see Fig. 3.1).[12] In expanding an earlier version of the data pipeline, Helena Sternkopf (2017) developed a data literacy maturity model. To describe 'individual levels' of data literacy maturity, this model adds 'assess and interpret' as a further category. For the purpose of evaluating organisations, the data literacy maturity model begins with two organisational aspects, 'data culture' and 'data ethics and security', which the 'Data Literacy Maturity Grid' records on four levels progressing from Uncertainty, Enlightenment and Certainty to Data Fluency (Sternkopf, 2017, p. 62f.).

These models do not promote any particular political orientation; they can be utilised for a range of goals. Critique of, or reflection on, the general texture of knowledge/power configurations *within* 'data' remains secondary. This specific understanding and practice of data literacy differs from conceptions within critical data studies, where data literacy primarily consists of a critical understanding of the societal function and impact of digital (big) data (Sander, 2020, p. 4).[13]

Taken by themselves, the website and these models suggest a neutral, relatively instrumental approach to developing skills, which might find use in relation to a range of questions, purposes or un/ethical endeavours.

Fig. 3.1 The data pipeline (https://schoolofdata.org/methodology [Accessed 2020-07-22])

However, as mentioned above, and as the next section foregrounds, School of Data aims at specific civil society organisations, activists, and related forms of democratic empowerment ('positive social change').

5 Countering Discrimination with Data

School of Data's cooperation with the non-profit organisation ufuq.de demonstrates the ways in which School of Data envision their role within data activism. Ufuq.de provides education programmes for civic education and prevention work in the fields of Islam, Islamophobia and Islamism. A brochure that resulted from a joint workshop held by the two organisations brings together the basic assumptions and technological skills that School of Data provides with the specific interests of ufuq.de and similar organisations relating to the documentation of and work on racist and other discriminatory right-wing incidents. The brochure, *Using Open Data in Anti-Discrimination Work. Approaches, Experiences, Pitfalls*

(Puvogel et al., 2017, henceforth *Using Open Data*), is available open access in German and English.[14]

The ideas of participation, civic engagement and (self-)empowerment through open data contained in *Using Open Data* echo the priorities outlined in Sect. 3 above: The brochure sketches the 'active citizen' as a citizen in a participatory relationship with their government, drawing information from a transparent state:

> When states make closed data transparent it can become public knowledge. In this way, civil society can strengthen its public control, demand accountability from political decision-makers and ultimately promote democratic processes. There is opportunity to actively shape society and participate in public debates. Open data, thus, serves as a tool to make the relationship between citizens and state institutions more transparent and participative. (Puvogel et al., 2017, p. 4)

Further, the brochure conceptualises open data as amenable to providing (numerical or visual) empirical arguments for campaigning, public relations and strategic decision-making in NGOs:

> Practically, the information can serve as a basis for argumentation, for progress reports or for strategic organisational decisions. One way that open data can help in campaign work is by communicating complex relationships in a more transparent and differentiated way. (Puvogel et al., 2017)

The examples *Using Open Data* contains, however, refer only in part to open *government* data, focusing also on (open) data generated by those people who are subject to discrimination.

One project mentioned in *Using Open Data* is the *Berliner Register* ('*Berlin Chronicles Against Racism and Right-Wing Extremism*'; henceforth *Berlin Chronicles*), which crowdsources data for the purpose of documenting and reporting on discriminatory incidents (Puvogel et al., 2017, p. 8). The *Berlin Chronicles* collect 'racist, anti-Semitic, LGBTIQ-phobic, anti-ziganist, right-wing extremist, right-wing populist and other incidents motivated by discrimination in Berlin's districts'.[15] These incidents are reported by citizens to *Berlin Chronicles* via the website, by email or telephone or in person at the drop-in centres city-wide. They are also reported by partner organisations, and collated from police reports, although prosecution is not required for an incident to be included in the statistics. The data set that emerges is both quantitative (how many incidents of what type in which part of the city, etc.) and qualitative (including narrative details of specific aspects of the incidents).

Alongside the human actors, multiple artefacts contribute to the *Berlin Chronicles'* activism, including flyers and information about the project, local drop-in centres, telephone lines, websites, encryption protocols for transmitting data or for PGP-encrypted emails, free software for the data analysis, visualisation and presentation in annual reports, and guidelines for police forces on categorising discriminatory incidents. The database itself functions as the essential connective between all the elements involved. The database becomes an agent in the potentially transformative processes that those people undergo who, for instance, use the database to relate their experiences of discrimination and assault to the similar experiences of others.

Since those affected by racist or other discriminatory practices are actively involved in the data collection process, this reporting is seen as an empowering practice (see also Gutiérrez, 2018, p. 143). In the foreground, these data practices appear similar to the use of hashtags such as '#metoo' to share and aggregate personal experiences of sexual harassment (Dolata & Schrape, 2018, p. 49). In the background, however, the forms for reporting discriminatory experiences as used in the *Berlin Chronicles* provide a much more structured dataset for further data analysis without erasing the unique aspects of each experience. Although the datasets incorporate reports by the police, the definitions and categories in use exceed those provided by the authorities. In this way, the simultaneous uncovering of individual *and* structural aspects of discrimination illuminates the specific role of data as a medium in the context of anti-discrimination work as highlighted by the *Using Open Data* brochure:

> The central roles of anti-discrimination work are: to make discrimination visible and to create a public sphere. This involves presenting individual cases of discrimination or assault anonymously, but, in particular, revealing the structural discrimination behind the individual experience. [...] Data makes it easier to visualise discrimination and its structural background. (Puvogel et al., 2017, p. 7)

Data are thus presented in the brochure as transcending purely quantitative operations and allowing for both quantitative *and* qualitative analysis. In this process, data literacy skills are helpful in refining and analysing data, recognising and visualising patterns of discrimination within datasets and, finally, communicating findings in a clear and plausible fashion (see Fig. 3.1 above). In this view, the criticisms of crowdsourced data as incomplete and not valid or representative do not necessarily identify a disadvantage of this form of data:

> […] raising awareness and empowering those affected […] often also includes the conscious decision to document subjective depictions that are not necessarily validated by other sources. In this way, reporting discrimination in itself becomes an act of self-empowerment by exposing the experience and making it visible. (Puvogel et al., 2017, p. 7)

This approach, described in the School of Data/ufuq.de materials, and put into action in projects such as the *Berlin Chronicles*, locates self-empowerment in the process of finding a data-based way of giving visibility to the experience of discrimination as part of structural discrimination. The empowerment of others via the reception, collation and anonymisation of crowdsourced data emerges here as a further important aspect of this type of data-based anti-discrimination work:

> The collection and documentation of discrimination can often be combined with practical advice and support for those affected. […] the [local drop-in] centres can also identify support structures and suggest individual ways of dealing with these issues. (Puvogel et al., 2017, p. 7)

This observation demonstrates the link between practices of data collection and documentation and social relationships revolving around communication, advice and support. The details of the data-based anti-discrimination work described in *Using Open Data* highlight the deeply collective, relational practices of doing data activism.

6 Political Subjectivities, Data Activism and Data Literacy

This final section brings together this worked example with the cultural and political theory presented in Sect. 2. How, specifically, do these data practices enact political subjectivity, and what broader implications do the theory and worked example have for data activism and data literacy?

In our analysis, we have identified all three forms of political subjectivity discussed in the theoretical literature. In the interviews, mission statements, websites, publications and data literacy models, School of Data enacts the idea of an individual (political) subject. The onus is on teaching data literacy as individual skills acquisition, it is on independent and self-determining data use, and on becoming fluent in using—but not necessarily critiquing the power of—data. When these skills are acquired, the materials suggest, individuals will be empowered to challenge today's widespread racist and other discriminatory practices.

However, with the shift from the data pipeline to the data literacy maturity model, School of Data also shifts away from a focus on individuals and towards a focus on organisations. The educational endeavour moves from individual skills acquisition to organisational development. This itself is an ambivalent reorientation. On one hand, it can seem like an elitist move to target organisations that are already active in activist practices, rather than anyone who walks in from the street. However, the organisations targeted by School of Data are themselves working within networks of grassroots activists and volunteers. If we understand non-profit activist organisations as collective configurations or communities of political practice, the move towards supporting organisational development enacts a shift from individual to collective political subjectivities. Data literacy is no longer seen as individual and context-free but as contextualised in collective projects.

The *Berlin Chronicles* show that inclusion in anti-discriminatory data practices can take many forms and occur on very different 'levels' of data literacy. These projects interlink several practices—and a range of human and non-human actors—to create a database of racist and other discriminatory attacks on which activist campaigns can draw. The *Berlin Chronicles* provides people experiencing or witnessing racism and discrimination with both a narrative and 'data points' on their experience. They may ***individually*** write an email or visit a drop-in centre (although they are equally as likely to go with a friend), but the key to the *Berlin Chronicles* is that each person's experience becomes part of a ***collective, distributed*** configuration, and their personal identity is ***obfuscated*** within this configuration. It is precisely through the anonymisation and aggregation of the large number of incidents that the data become a powerful force in anti-discrimination work that redistributes the sensible (Rancière) and makes the otherwise invisible visible.

Overall, thus, in our worked example, the explicit conceptualisations of citizenship and data literacy remain broadly within the liberal, individualistic state-citizen relationship of informing, debating, voting and individual empowerment. However, in the details of recent, specific, contextualised projects in which our focal organisations turn to action, we read an understanding of citizenship and data literacy which reach beyond this framing, and enact relational, collective and obfuscational political subjectivities. These projects potentially lead to empowerment by giving individuals anonymous access to resistant, counter-hegemonic practices and thereby entry into a collective, an assemblage of human and non-human agents, enacting distributed, networked data literacy.

The specific analysis suggests implications for data literacy. First, what would happen if data literacy models explicitly framed data literacy as a collective, relational endeavour? Would they have the same uptake in society as current models do? Would they have a different effect? Second, data literacy models or teaching would need to include a self-reflexive moment on how the data literacy model or educational approach itself subjectivates those becoming literate: as autonomous, self-determining, modernist individuals; as relational, collective beings; or as individuals empowered to obfuscate (and to turn the usually negative connotations of the word 'obfuscate' into a positive word for today's world in which 'we are data').

The example also suggests implications for data activism, which are perhaps best framed as open questions. Voice and bodily presence doubtless remain vital to activism. This said, we wonder whether the societal transformations potentially arising with datafication are also transforming political subjectivities beyond these classic forms of activism. What happens if we adopt a relational, collective approach to activism as distributed across multiple actors? How can datafying practices be thought beyond themselves, towards a more emancipatory perspective? Within dynamic data activist practices, involving humans, artefacts, symbols and texts, we recognise not only an opportunity for counter-hegemonic participation in public spaces and for the experience of self-transformation (speaking with de Lagasnerie: *désasujettissement or desubjugation*), but are also aware of the potential reinstatement of overarching, dominant cultural and discursive norms and orders, such as Enlightenment conceptions of political subjectivity.

Notes

1. Here, we follow the understanding provided by Dencik et al. (2019, p. 876): '[W]e should use data justice as a form of critique, a framework for shifting the entry-point and debate on data-related developments in a way that foregrounds social justice concerns and ongoing historical struggles against inequality, oppression and domination'.
2. In this paper, we will refer to School of Data Germany by 'School of Data'. Our analyses do not include material on School of Data in general, unless noted otherwise.
3. I01:5:7–8.
4. See https://okfn.de/en/themen/open_data/; or for a more precise definition: https://opendefinition.org/od/2.1/en/ [both accessed 2020-02-15].
5. Meanwhile text and site have been slightly changed: 'In a well-functioning, democratic society citizens need to know what their government is doing. To ensure this, transparency and democratic control, as well as free access to government data and information in addition to the ability to share this information with other citizens are needed. In the digital age, data is the key resource for social and economic activities. By opening up data, governments can help drive the creation of innovative businesses and services that can create social and economic value. The opening of data improves policy participation opportunities and facilitates informed involvement of citizens in policy making.' See https://okfn.de/en/themen/open_data/ [Accessed 2020-02-15].
6. See https://datenschule.de/team/ [Accessed 2020-06-29].
7. The interviews are referred to as I01 (the longer semi-structured expert-interview on School of Data in general and mostly educational matters from April 2018) and I02 (the shorter radio interview that was conducted before—March 2018—and live on air with a focus on the convergence of open data and anti-discriminatory work). I02's audio file is publicly available under a CC-BY-SA-NC licence: https://www.freie-radios.net/87814. The interviews were led in German language. Translations in this publication are ours.
8. 'It is not that they [the two developers in the team] only think of technical aspects: Of course, they also have a strong, in the sense of a strong political, opinion' (I01:8:24-25).
9. See https://datenschule.de/projekte/umweltdatenschule/ [Accessed 2020-06-29].
10. See https://datenschule.de/en/ [Accessed 2020-02-15].
11. The interviewee drops 'literacy' from the notion of 'Data Literacy Maturity', but refers prominently to it: 'wir haben das Data Maturity

Modell mit entwickelt, wo zivilgesellschaftliche Organisationen quasi die Möglichkeit bekommen, über ihre eigenen Datenfähigkeiten zu reflektieren.' (I01:15:30-32) English translation: 'We co-developed the data maturity model, with which organisations in civil society are enabled to reflect on their own data skills.'

12. See https://schoolofdata.org/methodology/ [Accessed 2020-02-16].
13. Carmi et al. (2020, p. 15), however, offer a framework for 'data citizenship' that tends to understand reactive and proactive aspects as complementary—'proactive' meaning to organise and engage in 'collective actions against various civic issues'. They conceptualise three areas: Data thinking, Data doing, and Data participation (Carmi et al., p. 10). Similarly, Wylie et al. (2019, p. 158f.) pair 'technological literacy' with an 'ethics of care'. Fotopoulou (2020), in her approach, addresses the intersection of data literacy with other literacies that appear meaningful for civil society organisations.
14. For the English version, see http://www.ufuq.de/Open_Data_english.pdf [Accessed 2020-06-21]. German version: https://www.ufuq.de/Open-Data-Antidiskriminierungsarbeit.pdf [Accessed 2020-06-21].
15. Cf. https://berliner-register.de/content/vorf%C3%A4lle-melden-report-incident [Accessed 2019-02-21].

References

Baack, S. (2015). Datafication and empowerment: How the open data movement re-articulates notions of democracy, participation, and journalism. *Big Data & Society, 2*(2), 1–11. https://doi.org/10.1177/2053951715594634

Baack, S. (2018). *Knowing what counts: How journalists and civic technologists use and imagine data*. Doctoral dissertation, University of Groningen. https://www.rug.nl/research/portal/en/publications/knowing-what-counts(4c94668a-c25c-43cb-9b36-5d54e3ff3c2e).html

Birchall, C. (2016). Shareveillance: Subjectivity between open and closed data. *Big Data & Society, 3*(2), 2053951716663965. https://doi.org/10.1177/2053951716663965

Brunton, F., & Nissenbaum, H. F. (2015). *Obfuscation. A user's guide for privacy and protest*. MIT Press.

Butler, J. (1993). *Bodies that matter. On the discursive limits of "sex"*. Routledge.

Butler, J. (2011). *Bodies in alliance and the politics of the street* [Lecture held in Venice, 7 September 2011, in the framework of the series The State of Things, organized by the Office for Contemporary Art Norway (OCA)]. http://www.eipcp.net/transversal/1011/butler/en

Butler, J. (2015). *Notes Toward a performative theory of assembly*. Harvard University Press.

Carmi, E., Yates, S. J., Lockley, E., & Pawluczuk, A. (2020). Data citizenship. Rethinking data literacy in the age of disinformation, misinformation, and malinformation. *Internet Policy Review, 9*(2) https://doi.org/10.14763/2020.2.1481

Charalabidis, Y., Zuiderwijk, A., Alexopoulos, C., Janssen, M., Lampoltshammer, T. J., & Ferro, E. (2018). *The world of open data: Concepts, methods, tools and experiences.* Springer International Publishing. https://doi.org/10.1007/978-3-319-90850-2

Cheney-Lippold, J. (2017). *We are data: Algorithms and the making of our digital selves.* New York University Press.

Couldry, N., & Mejias, U. A. (2019). *The costs of connection: How data is colonizing human life and appropriating it for capitalism.* Stanford University Press.

Couldry, N., & Powell, A. (2014). Big Data from the bottom up. *Big Data & Society, 1*(2), 2053951714539277. https://doi.org/10.1177/2053951714539277

Daly, A., Devitt, K., & Mann, M. (Eds.). (2019). *Good data. Theory on demand #29.* Institute of Network Cultures, The Netherlands.

Dander, V. (2014). How to gain knowledge when data are shared? Open government data from a media pedagogical perspective. *Seminar.net – International journal of media, technology and lifelong learning, 9*(2), 1–12. https://seminar.net/index.php/component/content/article/75-frontpage/current-issue/216-how-to-gain-knowledge-when-data-are-shared-open-government-data-from-a-media-pedagogical-perspective

Datenschule. (2021). Landing Page. Retrieved February 22, 2021, from https://datenschule.de/en/

de Lagasnerie, G. (2017). *The art of revolt: Snowden, Assange, Manning.* Stanford University Press.

Dencik, L., Hintz, A., Redden, J., & Treré, E. (2019). Exploring data justice: Conceptions, applications and directions. *Information, Communication & Society, 22*(7), 873–881. https://doi.org/10.1080/1369118X.2019.1606268

Dolata, U., & Schrape, J.-F. (2018). *Collectivity and power on the internet. A sociological perspective.* Springer.

Fotopoulou, A. (2020). Conceptualising critical data literacies for civil society organisations: Agency, care, and social responsibility. *Information, Communication & Society*, 1–18. https://doi.org/10.1080/1369118X.2020.1716041

Gleason, B., & von Gillern, S. (2018). Digital citizenship with social media: Participatory practices of teaching and learning in secondary education. *Journal of Educational Technology & Society, 21*(1), 200–212.

Gutiérrez, M. (2018). *Data activism and social change.* Palgrave Macmillan.

Gutiérrez, M., & Milan, S. (2017). Technopolitics in age of big data. In F. Sierra Caballero & T. Gravante (Eds.), *Networks, movements & technopolitics in Latin America: Critical analysis and current challenges* (pp. 95–109). Palgrave Macmillan.

Hahn, H., Helm, J., Neujeffski, M., & Perseke, K. (2017). *Jugendarbeit im digitalen Wandel. Kompetenzen, Daten und digitale Tools.* Open Knowledge Foundation Deutschland e.V. / Projekt: Demokratielabore. https://bedarfsanalyse.demokratielabore.de/

Hintz, A., Dencik, L., & Wahl-Jorgensen, K. (2018). *Digital citizenship in a datafied society.* Polity Press.

Isin, E. F., & Ruppert, E. (2015). *Being digital citizens.* Rowman & Littlefield.

ISTE – International Society for Technology in Education (ISTE). (2019). *Digital citizenship in education.* https://www.iste.org/learn/digital-citizenship

Kennedy, H. (2018). Living with data: Aligning data studies and data activism through a focus on everyday experiences of datafication. *Krisis. Journal for Contemporary Philosophy.* https://krisis.eu/living-with-data/

Kessl, F. (2005). *Der Gebrauch der Eigenen Kräfte: Eine Gouvernementalität Sozialer Arbeit.* Juventa.

Macgilchrist, F. (2019). Cruel optimism in edtech: When the digital data practices of educational technology providers inadvertently hinder educational equity. *Learning, Media and Technology, 44*(1), 77–86. https://doi.org/10.1080/17439884.2018.1556217

Mack, E., & Gaus, G. F. (2004). Classical liberalism and libertarianism: The liberty tradition. In G. F. Gaus & C. Kukathas (Eds.), *Handbook of political theory* (pp. 115–130). SAGE.

Milan, S., & Treré, E. (2019). Big data from the South(s): Beyond data universalism. *Television & New Media, 20*(4), 319–335.

Milan, S., & van der Velden, L. (2016). The alternative epistemologies of data activism. *Digital Culture & Society, 2*(2).

Open Knowledge Foundation Germany. (2019): Offene Daten. Website. Retrieved May 20, 2019, from https://okfn.de/en/themen/offene-daten/

Puvogel, M., Neujeffski, M., & Nordbruch, G. (2017). *Using open data in antidiscrimination work.* ufuq.de. http://www.ufuq.de/Open_Data_english.pdf

Rancière, J. (1999). *Disagreement: Politics and philosophy.* University of Minnesota Press.

Rancière, J. (2001). Ten theses on politics. *Theory & Event, 5*(3) https://doi.org/10.1353/tae.2001.0028

Rancière, J. (2013). *The politics of aesthetics: The distribution of the sensible* (G. Rockhill, Trans.). Bloomsbury Academic.

Ratner, H., & Ruppert, E. (2019). Producing and projecting data: Aesthetic practices of government data portals. *Big Data & Society, 6*(2), 1–16.

Rawls, J. (1996). *Political liberalism.* Columbia University Press.

Ribble, M., & Bailey, G. D. (2007). *Digital citizenship in schools* (1st ed.). International Society for Technology in Education.

Sander, I. (2020). What is critical big data literacy and how can it be implemented? *Internet Policy Review, 9*(2) https://policyreview.info/articles/analysis/what-critical-big-data-literacy-and-how-can-it-be-implemented

Saward, M. (2006). Democracy and citizenship: Expanding domains. In J. S. Dryzek, B. Honig, & A. Phillips (Eds.), *The Oxford handbook of political theory* (pp. 400–420). Oxford University Press.

Schaal, G. S., & Heidenreich, F. (2006). *Einführung in die politischen Theorien der Moderne*. Budrich.

School of Data. (2021). Landing Page. Retrieved February 2, 2021, from https://schoolofdata.org/

Sternkopf, H. (2017). *Doing good with data. Development of a maturity model for data literacy in non-governmental organizations.* Master's Thesis, Hochschule für Wirtschaft und Recht Berlin. https://datenschule.de/files/workshops/DoingGoodwithData-DevelopmentofaMaturityModelforDataLiteracyinNGOs_HelenaSternkopf.pdf

Wylie, C., Neeley, K., & Ferguson, S. (2019). Beyond technological literacy. Open data as active democratic engagement? In R. Reichert, K. Wenz, P. Abend, M. Fuchs, & A. Richterich (Eds.), *Digital Culture & Society 4(2)* (pp. 157–181). Transcript.

Zuboff, S. (2019). *The age of surveillance capitalism: The fight for a human future at the new frontier of power*. PublicAffairs.

CHAPTER 4

Sexist Hate Speech as Subjectivation: Challenges in Media Education

Britta Hoffarth

Abstract This chapter is dedicated to exploring the practice of sexism—in the sense of sexist hate speech—in digital media and its discursive relationship to theories of subjectivation and education. While I do not approach the issue via the concept of discourse in a formal analytical sense, I reference a Foucauldian view of language drawing on theories of discourse and identifying language and speaking as instruments of power and knowledge. After surveying the current state of digitisation and media education, I will use examples of sexist hate speech to examine the relevance of the gendered orders in force in media and beyond and illuminate a gap in theories of media education in terms of their neglect of the analysis of power relations.

Keywords Sexist hate speech • Online • Subjectivation • Media education

---

B. Hoffarth (✉)
Institut für Erziehungswissenschaft, Universität Hildesheim, Hildesheim, Germany
e-mail: hoffarth@uni-hildesheim.de

P. Bettinger (ed.), *Educational Perspectives on Mediality and Subjectivation*, Palgrave Studies in Educational Media,
https://doi.org/10.1007/978-3-030-84343-4_4

1 Digitisation and Issues of Media Education

Techno-sociological discourses are now referring to digital media as the fourth narcissistic affront to "anthropocentric humanism" (Deuber-Mankowsky, 2007, p. 278) and pointing to a "border between the human and the technological that has become fragile" (Eickelmann, 2017, p. 19).[1] At the outset of the internet age around the turn of the millennium, it was still challenging to adequately evaluate these new media of digital communication against the backdrop of established media cultures, such as television and print media. The challenge today, by contrast, appears to lie in examining the specifics of digital communication beyond its everydayness, its ubiquity and therefore, to an extent, its invisibility, read in the frame of Bourdieu's concept of the doxa.[2] However, this effect of the digital space becoming 'invisible', in the sense of its de-thematisation and its disappearance into the self-evident, only holds for those who do not struggle to disseminate their own content, who have access to technology, stylistic devices, and hegemonic forms of representation. In this context, the term 'digitisation' describes not only the quantitative spread of digital technologies, but also a node in the discourse of technological progress, which at the same time stands for forms of social and cultural transformation without being able to explicate them. It is a fundamental paradox of social media that the conception of access to the internet, both as a knowledge resource and as a medium of self-representation, as a basic human right stands alongside the simultaneous symbolic regulation of participation in digital spaces, as is apparent in hate speech and other phenomena.

Within these to a degree contradictory considerations and alongside current studies on media use (Bertelsmann Stiftung, 2019; Capurro, 2017; Dengel, 2018; Kaspar et al., 2017; Kerres, 2018), current considerations from the educational field on the significance of digital media to adolescents engage with both the risks and the potential of digital practices. It is a discussion that emerges on at least two levels: in the continuation of established discourses in media education, and in education studies' novel centring of media in the consideration of matters around political education. While much discussion of digitisation in pedagogy relates to matters of media literacy (Gesellschaft für Medienpädagogik und Kommunikationskultur in der Bundesrepublik Deutschland e. V. (GMK), 2013; Mitgutsch, 2009), I propose to take a distinct approach, exploring the phenomenon in terms of education theory, with an emphasis on social critique and the analysis of power relations.

The exploration of facets of education theory (*Bildungstheorie*)[3] within media education, as inspired by the discipline of media studies, is a relatively recent phenomenon. The reverse preoccupation, of education theory with matters of media has a tradition; this said, media and their associated cultures remain fairly marginal in general education studies (Sesink, 2014). The task of addressing the relationship between media and education seems all the more urgent for the current development of the new and constantly renewing media into a "trouble spot", as Sesink (ibid., p. 12) puts it. Referring to Marotzki and Jörissen, Iske proposes to systematically meet current challenges with a "structural media education"[4] that is not based on representational media, but on the phenomenon of 'mediality', which abstracts from media phenomena and media types and focuses on overarching aspects of form and structure (2014, p. 4). This definition conceives of mediality as an anthropological moment, as a constitutive "(a) prerequisite for symbolism, (b) structural condition of concrete (cultural-historical) forms of articulation and thus (c) as a structural condition in education theory for the construction of relationships with the self and with the world" (Jörissen, 2014, p. 503). Iske, who identifies Humboldt's theory of education as the starting point of this approach to media education, notes the difficulty of determining the extent to which "the conventional understanding of education has to be reformulated in view of changes in social conditions" (2014, p. 7). We might formulate this issue, and the theoretical ambition inherent to it, even more radically thus: While we may safely assume that translations of neo-humanist concepts of education already exist, it seems necessary to determine more precisely what the concept of 'social conditions' actually refers to empirically, in order to access the (technical) historical and socio-structural contexts of reflection on education, specifically media education. Gendered orders and their immanent violence are one facet of social conditions which this chapter will explore, aiming to illuminate their significance for those growing up in both gendered and digitised societal contexts.

While they take traditional educational concepts from the Enlightenment into account and thus contrast approaches to media literacy (a set of acquirable skills)with an idea of education that calls into question overly linear ideas of the transmission of knowledge, current concepts of media education could be accused of implying a relatively unproblematic notion of education. We may concur with Alfred Schäfer in essentially assuming that education does not occur as a process of linear transformation, but instead is accompanied by various disruptions (Schäfer, 2019) that are, as we will go on to note, closely linked to the problem of representation. It appears, therefore, that we are in need of a concept of media education that considers both the political implications of social differences in how individuals engage with media and the fragility of processes both of education and of the reception of media content.

This chapter will explore the phenomenon of sexist online hate speech in the light of feminist media studies, illustrating the various facets of gender-based violence in online environments and its inherence to these settings. It will subsequently proceed to develop a theoretical framework for examining the phenomenon, discussing the relationship between representation and subjectivation. Concluding, I will consider this framework's productivity for rethinking issues of media education.

2 Sexist Hate Speech Online

Sexist discrimination encompasses structural and individual practices of discrimination on the basis of gender. This discrimination does not, as the term might suggest, refer only to the topos of sex, but can occur towards members of both the female and the male sex; a diversity, persistence and sharpness, however, is distinctly observable in sexist discrimination towards femininity: "everyday sexism exemplifies male entitlement" (Jane, 2017a, p. 1). Döring and Mohseni (2018) note that current studies on male domination in social media demonstrate the general abundance of negative and hateful feedback in online communication and that female participants in particular face substantially greater volumes of negative and particularly sexualising and cruel comments on their appearance than do male participants (ibid., p. 511). We can cite Eickelmann in defining internet hate speech as a form of "mediatised" (2017, p. 21) disregard, including defamation, insults, death and rape threats, which function as a vengeful or controlling imposition of impending violence and a punitive sanction applied to public media appearances marked as feminist or even just female. In both the mass media and social convergence media with user-generated content such as YouTube and Whatsapp, media stagings of femininity appear to be particularly conflictual in their susceptibility to social sanction; for example, young girls who send nude photos of themselves in the course of so-called sexting (cf. Bedor, 2015; Amundsen, 2019; Döring, 2014) may suffer severe loss of social reputation if they are subsequently the subject of a 'revenge porn' attack, and suicides have been documented in this context (cf. Eickelmann, 2017). The dilemma facing female adolescents here consists in the contradictory social demands imposed on them; the expectation of self-performance of sexual attractiveness clashes with the requirement of modesty and self-restraint and the oftenly punitive response to any self-confident or self-determined bodily displays. Victim-blaming is closely linked to the experience of online harassment (cf. Lumsden & Morgan, 2017). There appears here an interpretation of girls' sexual self-determination as a threat to a socially desirable concept of femininity. In other words, sexist bullying or hate speech in digital media always refer back to power relations that already possess social significance outside media spaces. The social sanction falls upon girls' offensive (in both senses of the word) presentation of their sexuality, not the illegal publication of the images. The common variants of sexist hate speech online, such as threats of rape or death and doxxing (the publication of personal data such as the victim's home address), are both individual attacks and massively restrictive of users in their opportunities to participate in the digital social sphere, their sense of autonomy, their identity, dignity and wellbeing (cf. Jane, 2017a). Systematic campaigns of harassment, as in the #GamerGate debate, see coordinated attacks launched on female users who are 'particularly visible' in specific internet communities (cf. Eickelmann, 2017). Jane perceives the existence of a gendered 'digital divide' (cf. 2017b).

I wish to propose an understanding of hate practices in this context as political practices systematically related to the devaluation of the female. The concept of the political in this understanding cannot be reduced to questions of government; it draws much more on active and influential participation in the social sphere (cf. Celikates & Gosepath, 2013). Such an approach encompasses two key aspects: First, it identifies hate speech not as a random, individual phenomenon, but rather as fed by socially established gendered orders and the misogyny enshrined in them. Second, the political nature of the practices examined here relates to their revolving around a struggle to assert public spheres and visibility and ultimately a discursive struggle to assert truth. In the discriminatory invocation (see section below), the user attacked experiences a de-rationalisation and a loss of authority through gendering and sexualisation; this denormalises women's participation in digital public spheres and beyond, mainly due to its restriction of spaces for self-definition and autonomous action. We may also perceive this de-normalisation of participation as a latent strategy of disempowerment of women, and specifically of women who identify as feminists or queer feminists. In its disruption of the idea of femininity, a feminist and implicitly queer critique of identity thus works on the three levels of normalisation, empowerment and deconstruction, which will be important later on in my proposal of a form of media education that responds to the phenomenon of hate speech on a range of levels.

3 The Negativity of the Medium: Representation Is (Not) the Problem

Representation is at the heart of media. One illustration of this which is particularly pertinent to matters of sexism is the so-called Bechdel test, or "The Rule", which originated from the comic strip *Dykes to Watch Out For* by Alison Bechdel (cf. 1985). Two protagonists of the comic talk about the quality of cinematic experience and the question of when it is worthwhile for them to go to the cinema: "I only go to a movie if it satisfies three basic requirements. One, it has to have at least two women in it... who, two, talk to each other about, three, something besides a man."[5] Until just five years ago, about 90% of the films made in Hollywood failed the test; the situation is currently changing.[6] The test is of interest to this discussion because, in spite of its charmingly pragmatic productivity, its ability to easily illustrate the problem of representation, and thus its heuristic problematisation of visibilities and gendered orders, it, or the critique of representation implicit in it, lacks an essential moment of mediality. This type of critique of representation implies that there are 'correct' forms (or quantities) of representation and undertakes a simplification and a problematic naturalisation of the relationship between sign and signified. Mersch (2006), however, considers the definition of representation as 'successful' to be fundamentally problematic: "There are media because there is alterity. Otherness means an 'other' that initially refuses access, that needs a third party to guarantee its mediation, its symbolisation, storage, transmission or communication" (ibid.).

This points to the necessity of fundamental reflection on the constitutive difference between sign and signified and thus on the relationship between representation in media and what is represented, a relationship which, as I will now discuss using terms proposed by Theodor W. Adorno, appears essentially characterised by fragility. In Adorno's view, critique forms itself as a relationship of 'non-identity'. In founding the idea of the "the subject-object dichotomy" (Adorno, 2004, p. 6) he articulates a questioning of the "gesture of domination through identification"

(Schäfer, 2004, p. 92) and assumes that designations (as practices of signifying), as appropriations by a symbolic order, always already represent a violent act. Adorno assumes that neither things nor the subject's self can be transparent to the subject. The I, in Adorno's terms, can never be the object of its own comprehensive knowledge; a term can never mean the object in itself nor can it be appropriated through understanding. Schäfer observes: "In contrast to such thinking that aims at conceptual and technical control over the world, that assumes that the world has to submit to subjective knowledge, Adorno's counter-perspective emphasises the 'primacy of the object'" (ibid.). Adorno's dialectic thus aims at reflecting on the contradictory circumstance that the object can never be identical with the concept it is identified with (ibid.), but is always more, means more, is—to speak with Laclau et al. (2012)—overdetermined.[7] In Schäfer's view, Adorno illuminates in this way the "resistant nature of the object, its enigmatic character which eludes any identifying ordinance [*Verfügung*]" (Schäfer, 2019, p. 119), which implicates the possibility of education. "In this context, this experience of one's own self [which emerges from it, B.H.] is that of the failure of a subject figure exerting control" (ibid.). Drawing on Adorno, we could comprehend the fundamental, characteristic property of media as a "disturbance" (Schäfer & Thompson, 2010, p. 141) which embodies in a fundamental manner the relationship of the subject to itself and to the world. What the medium, the text, the image shows (indeed generates) is always simultaneously incomplete and more than what is shown. Representation thus never leads to a recognition of the object; the processes of representation and reception nevertheless remain productive, but, in terms of a theoretical formulation of the mediality of technical apparatuses, the effort to produce an adequate representation always fails. Mediality, as Mersch (2006) points out, appears fundamentally characterised by negativity. Following this view, and in light of a critique of representation, we might perceive the medium itself as a third party, a figure of alterity and also of negativity, that is always present, and in its presence erases itself. This may point the way analytically, opening up a concept of media (Mersch, 2006) by highlighting the constitutive 'media-tion' of sociality and subjectivity, which means, effectively, that there is no practice outside media. Likewise, the converse holds, in that practice taking place within a technical medium, such as internet communication, is always a practice of the social. This concept of mediality thus calls for research that does not exclude or overlook any aspect of the social and whose leading focus is not a specific object in the sense of a technical device, but rather a perspective on cultural practices. This understanding of media critique and analysis gives rise to a systematic approach whose chief concern is to formulate questions around these arrangements of technologies and cultural practices as they emerge, for example, in the phenomenon of online hate speech. To come full circle at the conclusion to this section, we might reflect that representation is not the sole category at the heart of media. Krämer puts it much more precisely—mediality can be understood as performativity (Krämer, 2002, p. 329).

4 Approaches to Subjectivation

Having discussed the problem of representation with the help of Adorno's critique of the assertion of identity, the section that follows will explore the phenomenon of sexist hate speech from the perspective of subjectivation, in order to identify the production of relationships with the self and with the world by hate speech in a context of the analysis of power. The subject as defined by Enlightenment and idealist thought appears to us as autonomous, sovereign and capable of reason (cf. Reckwitz, 2008, p. 75); a post-structuralist conception instead emphasises its fundamental decentrality (cf. Zima, 2000), pinpointing subjectivity as arising not from a core within the individual, but rather in a fundamental reference to another. Michel Foucault (2000, p. 240) considers subjectivation in modernity to occur as a form of government in three forms of objectivation: "Forms of knowledge, power technologies and processes of self-formation" (Lemke, 2006, p. 269). By government, Foucault does not refer to the power of leadership invested in a sovereign, an autocratic authority, but rather to the decentralisation of leadership, controlled by a complex network of social forces and forms of knowledge (Pieper & Gutiérrez Rodríguez, 2003, p. 10), and to leadership of self-leadership. How we learn to lead ourselves is always closely linked to how we understand ourselves, and as whom; analytical observation in this context is primarily directed towards the moment of "self-deciphering" (Lemke, 2004, p. 161). This leading of ourselves is interconnected with being led: Subjectivation unfolds in a dual, dialectically interlocking movement of 'being called' and obeying the call, as outlined by Althusser:

> I shall then suggest that ideology 'acts' or 'functions' in such a way that it 'recruits' subjects among the individuals (it recruits them all), or 'transforms' the individuals into subjects (it transforms them all) by that very

> precise operation which I have called interpellation or hailing, and which can be imagined along the lines of the most commonplace everyday police (or other) hailing: 'Hey, you there!' Assuming that the theoretical scene I have imagined takes place in the street, the hailed individual will turn round. By this mere 180-degree physical conversion, he becomes a subject. Why? Because he has recognized that the hail was 'really' addressed to him, and that 'it was really him who was hailed' (and not someone else). (Althusser, 2014, p. 264)

In Althusser's study *Ideology and Ideological State Apparatuses*, interpellation denotes a process in which ideology in the form of an address, the "Hey, you there!" to which the person called turns, makes individuals into subjects. Judith Butler takes up the concept of the Althusserian interpellation in noting the subject's dependence on the conditions of being that precede it, relating this dependence to an analysis of power relations: "Subjection" signifies the process of becoming subordinated by power as well as the process of becoming a subject. Whether by interpellation, in Althusser's sense, or by discursive productivity, in Foucault's, the subject is initiated through a primary submission to power (Butler, 1997, p. 2). To Butler, however, subjection/subjectivation does not happen without the moment of foundation, of inauguration into an non-establishable position (ibid., 2001, pp. 9–10).

In *Excitable Speech* (2006), Butler assumes, with reference to Austin and Althusser, that language is performative; that is, the hurtful power of insults arises because language has an effect beyond speaking in its quoting of social 'truths' and thus identifies the subjects thus addressed in a way that, first, misjudges them and, second, limits their power of action. The question of validity—that is, when something is considered hate speech—is not amenable to a definitive answer. Rather, from a point of view drawing on the theory of performativity, I presuppose the incommensurable productivity of speaking and identify the key question in this regard as relating to the way in which speaking becomes productive.

In this context, we might denote the influence of acts of addressing/hailing on relationships with the self and with the world as a point of connection between education theory and the theory of subjectivation. While both Ricken (1999) and Koller (2011), for example, propose a reading of the concept of subjectivation as an educational concept, we would consider it legitimate here to distinguish education in the sense of education and subjectivation as drawn from the thought of Foucault and Butler.

Foucault's critique of modernity's techniques of individualisation traces the subject as an object measured in specific rationalities; Foucault describes the process of subjectivation as a dual movement of becoming in and subjugation to social norms. Butler further elaborates how this becoming/subjugation inscribes itself in the subject, forming the "trace of the sociality of the subject in the subject" (Bierbaum, 2004, p. 184). Neither Foucault nor Butler conceive of subjectivation-as-education as the outcome of conditions of oppression. They rather regard power and the subject as intertwined and not juxtaposed. In Butler's view, norms are precisely the authority that constitutes the subject—but again, they do not determine it. Education or self-education therefore cannot lead to self-liberation, since freedom can be understood as permanently embedded in power relations.

I intend now to bring the specific conditions of becoming a subject—in relation to gender orders—into view by empirically operationalising the dialectical relationship between self-techniques, invocations/addresses and figurations of knowledge via the concept of addressing proposed by Reh and Ricken (2012), which emphasises subjectivation's positioning effect. The authors understand a positioning as a dual relational event: first, a setting in relation to oneself, others and the world, initialised by the act of addressing, and second, a setting in relation to this setting-in-relation (cf. Reh & Ricken, 2012, pp. 39): "Subjectivation therefore refers to that practical (and precisely not, or not simply, reflective) process of engagement with culturally presented subject forms in which the individual as a self makes itself as well as being made a self by others; this engagement encompasses both subjection to the social orders in force and transgressions of them, revealing both consolidations and shifts [in these orders]" (ibid., p. 40).

For the targets of sexist practices of addressing, subjectivation in this sense does not multiply their possibilities of signification, action or positioning, but instead brings the play of identities to a standstill and fixates these identities on an insurmountable otherness. The sexist-identifying interpellation produces positions by imposing onto individuals a knowledge (about themselves and others) that stems from a specific order of knowledge and suggests certain self-techniques while preventing others. Being addressed in a sexist way that violates my need for self-determination, I am required to relate to an order of knowledge that marginalises and subordinates me.

In Reh's and Ricken's terms, the invocation/interpellation, the calling that calls upon the subject to position itself, is a specific practical act. Acts of addressing, such as insults, represent practices which establish a reference to a person linguistically and physically, through acts of being touched, being looked at, gestures, and so on. In so doing, they reproduce facets of social orders and the relativity of the participants in the address; the performative double movement that ensues acts both to suspend these participants' subject positions and to confer validity upon them while shifting them. Reh and Ricken propose a three-step heuristic that could empirically examine processes of subjectivation via practices of addressing and re-addressing. They first seek to identify the "normative horizons" (ibid., p. 44) within which acts of addressing acquire meaning and therefore validity; second, they interrogate the positioning, anticipation and setting of relations; and third, illuminate the emergent spaces of possibility, the possibilities of a second-degree putting-into-relation, so to speak, in which the subject interprets the positioning performed in the address (cf. ibid., pp. 44). In this way, they unfold a methodical approach to examining practices of addressing, which takes into account the entanglement of the interactional and discursive levels.

5 Online Hate Speech as Subjectivation and the Challenges of Media Education

The section that follows will provide an analytical view of online hate speech in reference to the theoretical framework outlined above, before proceeding to discuss the findings of this analysis and their significance for an attempt to rethink media education.

Research based on the theory of subjectivation as set out by Butler has formulated an interest in the conditions under which an individual becomes a subject. This perspective both takes account of social structures and acknowledges a concept of practices that emphasises not the intentionality of action (and thus the autonomy of the subject), but rather its performative effects. Proceeding from the idea of negativity in Adorno and the subsequent concept of the performativity of the medium, we identify the character of mediality not as an aspect of technological artefacts, but rather as a fundamental moment of thinking and speaking (cf. Bernardy, 2013, p. 113). Applying this insight to online hate speech alongside the framework of recognition theory, we find ourselves facing the question of the productivity of speaking on the internet, as explored through the concept of subjectivation earlier in the chapter. The approaches to the phenomenon mentioned here demonstrate that we cannot think pedagogically about these matters if we do not incorporate an analysis of power relations into our perspective.

In connection with media and subjectivation, Jörissen speaks, for example, of the "constitutivity of mediality for relationships with the self and with the world" (Jörissen, 2011, p. 12), thus assuming that relationships involving the self, such as those of the self to the world and to others, and their transformations are fundamentally characterised by the moment of mediality, emerging in the field of tension between representation and productivity. Considering power relations in the light of theories of subjectivation complements the concept of mediality in this context by seeking to identify fundamental limitations of the spaces of possibility in which transformations of relationships with the self and with the world can take place.

An analysis of hate speech founded on approaches to power relations would perceive it as the expression and initiation of multi-dimensional social processes of education—'education' being understood here not as a process of the imparting and acquisition of knowledge, but in the sense of a transformation of self- and world relations. Acts of hate speech demand a response to the addresses they make, while requiring subjects to relate to the confrontational address in a way that also makes them relate to the world as their social context. Hate speech already includes a certain interpretation of the social which it imposes on the addressed subject. Speaking functions in this context via the citation of a social order: the word "slut" or "bitch" only hurts when the addressee knows about its degrading meaning and the structures that produce that meaning. Put differently, being hurt by such an address is an experience of the implicit violence of being identified in the context of a hegemony of meanings. To reference Reh and Ricken (2012), the "normative horizons" (ibid., p. 44) within which acts of addressing acquire meaning and therefore validity produce a kind of positioning, an anticipation and setting of relations, which lead to the emergence of spaces of (im)possibility. 'Education' in this context means actively relating to the claims of truth implemented in hate speech articulations: the truth of the social which includes the truth of the inferiority of the person being addressed. Approaches to media education that reflect these aspects of interconnection between subjectivation and education through online hate speech transcend a normative concept of media literacy (which simply aims to delineate the *right* way of using a specific technology) and prompt their recognition as political education The specificity of sexist violence in social media in contrast to 'analogue' violence lies in the peculiarities of technological communication and the cultural practices emerging from them: the blurring of boundaries between asynchronicity and synchronicity of communication, the physical absence and (usually) anonymity of the perpetrator, the possibility of duplication and dissemination of data and of the initiation of hate campaigns, and the difficult traceability and confusion of routes and spaces communication on the internet. In the digital world, the public sphere unfolds across heterogeneous contexts, which may protect the perpetrators and increase the exposure of those affected. Understanding of and reflection upon these facets of violent online sexism may cast light on the nature of the apparently personal (of the hate-speech insult) as political.

In the context of this investigation, the concept of mediality refers to the field of tension between representation/mediation and the contingent production of meaning as attributable not solely to the so-called media, but to every materiality or phenomenon that intervenes in the social in this way (representing, and simultaneously rupturing its productivity in its representativeness). Elsewhere, I have proposed linking this concept with Butler's notion of the performative (cf. Hoffarth, 2009): speaking of the performativity of (mass) media makes it possible to perceive at their potentially deconstructive productivity. Digital media place the interactive action of the user at their centre and thus add the practice of inscribing to the practice of (productive) reading of media text as identified in earlier work from a cultural studies perspective. Led by this insight, we can observe that the users of online media communicatively shape the digital space. The medium's tendency to disappear in communication—its negativity—, which contributes to the assertion of representation as an unbroken relationship between sign and signified, shores up the power of the violation inherent in hate speech. "Due to their technical-symbolic constitution, media architectures, in comparison to material settings, provide more variable, more dynamic, often more complex and, not least, often significantly novel variants of such 'performative' structures, which are accordingly accompanied by new potential for reflection" (Jörissen, 2011, p. 12)—and, I would add, new potential for instances of violation.

The constitutively violent nature of the acts of addressing in sexist hate speech consists not only in the hurtful invocation, not only in the use of the insult itself, but also in the quasi-dual performativity of the invocation, in which the addressee is called upon to relate to the insult, to interpret it. This means that a concept of media education with the capacity to lay bare these consequences for subjectivation would need to take into account the representational ruptures in the media lives of those we wish to engage in this education. Jörissen's proposal is to incorporate the concept of consciousness or awareness of mediality into the conception of media education (Jörissen, 2011, p. 72). The development of this type of mediality awareness would possess a multi-faceted pertinence both to those exposed to sexist violations and to those who inflict these injuries. Measures aimed at preventing violent behaviour on the internet should address the latter in particular, rather than primarily seeking to limit young female users' scope of action in response to 'what might happen' to them. This concept of media education should also include political strategies drawing on anti-discrimination pedagogy, focusing on empowerment alongside reflection on privilege.

A useful concept for illuminating the strain on an approach to media education motivated by social and subject theory and attempting to address discriminatory structures is that of the "trilemma of inclusion" put forward by Mai-Anh Boger (2019), which consists in three dilemmas between the concepts of deconstruction, normalisation and empowerment, which it defines as constitutive moments of inclusion and which I linked to the phenomenon of hate speech above. Only two of these concepts can ever 'go' together, while the third must logically be excluded—and yet all three constitute the practice of inclusion which we perceive here as a productive idea for political education and a critique of exclusion. While empowerment strategies function at both an essentialist and an individual level, strategies of normalisation, resting on an assumption of integration of someone into an assumingly normal assemblage, pursue the goal of participation in collective normalities. Deconstruction, finally, takes place in the form of analysis and critique of these normalities and essentialisations.[8] A media education inspired by subjectivation theory, then, would aim to critique both social and technological structures alongside practices of representation. The challenge regarding the trilemma especially consists of analysing the meanings of normalisation, empowerment and deconstruction, that is, the desire for identity, empowerment and non-identity, in their various combinations in relation to social hegemonies such as gender, race and class, and of taking these meanings into account when designing media-education. At the same time, this pedagogy would need to consider the contingency of subjectivation and education as transformations of relationships with the self and with the world. Media education, in this sense, means using the productive tension in the trilemma to analytically examine and pedagogically work with transformations in individuals, institutions, and power relations in order to prevent hate speech-related violations on the internet.

Notes

1. Most quotations were originally published in German. All translations of these quotations were provided by the author.
2. Bourdieu did not have digital media in mind when he adopted the concept of doxa from Greek philosophy (cf. Fröhlich & Rehbein, 2008, p. 79). The principal facet of the concept, as used by Bourdieu, that meets a crucial moment of digital media is its compatibility with postmodern conditions and its becoming 'natural' or normal and thus self-evident in its everyday use.
3. Cf., for example, Wischmann, 2018, on the German discourse on *Bildung* as education.
4. What follows will use the concept of 'education' to mark processes of transformation of self- and self/world relations (in contrast to practices of parenting or processes of socialisation and in contrast to education as pedagogy).
5. Alison Bechdel, ‚The Rule', *Dykes to Watch Out For*, https://dykestowatchoutfor.com/wp-content/uploads/2014/05/The-Rule-cleaned-up.jpg, 19.05.2020.
6. https://bechdeltest.com/, 20.02.2020.
7. Bernardy points out that it is not necessarily contradictory to suggest an argumentative proximity between Adorno's concept of the non-identical and the deconstructivist idea of mediality (Bernardy, 2013, p. 113).
8. In terms of the theory of science, the concept is simultaneously a provocative and integrative project, in that it takes up divergent, even contradictory, theories and thus challenges scientific thinking and argumentation at various levels.

References

Adorno, T. W. (2004). *Negative dialectics*. Taylor & Francis e-Library.

Althusser, L. (2014). *On the reproduction of capitalism: Ideology and ideological state apparatuses*. Verso.

Amundsen, R. (2019). "Kind of like making porn of yourself": Understanding sexting through pornography. *Feminist Media Studies, 19*(4), 479–494.

Bedor, E. (2015). Sexting panic: Rethinking criminalization, privacy, and consent. *Feminist Media Studies, 15*(6), 1065–1066.

Bernardy, J. (2013). Formen und Figuren literarischer Artikulation bei Foucault und Wittgenstein. In S. Niklas & M. Roussel (Eds.), *Morphomata: Vol. 11. Formen der Artikulation. Philosophische Beiträge zu einem kulturwissenschaftlichen Grundbegriff* (pp. 109–141). Fink.

Bertelsmann Stiftung. (2019). *Monitor Digitale Bildung: Berufliche Ausbildung im digitalen Zeitalter.*

Bierbaum, H. (2004). Kritische Theorien des Subjekts (und) der Bildung: Foucault/Butler und Heydorn/Koneffke zwischen Differenz und Annäherung. In L. Pongratz, W. Nieke, & J. Masschelein (Eds.), *Schriftenreihe der Kommission Bildungs- und Erziehungsphilosophie der DGfE. Kritik der Pädagogik – Pädagogik als Kritik* (pp. 180–199). Leske u. Budrich.

Boger, M.-A. (2019). *Theorien der Inklusion: Die Theorie der trilemmatischen Inklusion zum Mitdenken* (1. Auflage). Münster: Edition Assemblage.

Butler, J. (1997). *The psychic life of power: Theories in subjection.* Stanford University Press.

Butler, J. (2001). *Psyche der Macht: Das Subjekt der Unterwerfung* (1st ed.). *Gender studies: 1744 = n.F. 744.* Frankfurt am Main: Suhrkamp.

Butler, J. (2006). *Hass spricht: Zur Politik des Performativen.* Suhrkamp.

Capurro, R. (2017). *Homo Digitalis.* Springer Fachmedien Wiesbaden.

Celikates, R., & Gosepath, S. (2013). *Politische Philosophie. Reclams Universal-Bibliothek: Vol. 18473.* Reclam.

Dengel, A. (2018). Digitale Bildung: ein interdisziplinäres Verständnis zwischen Medienpädagogik und Informatik. *MedienPädagogik: Zeitschrift für Theorie und Praxis der Medienbildung, 33*(0), 11–26. https://www.medienpaed.com/article/view/558

Deuber-Mankowsky, A. (2007). *Praktiken der Illusion: Kant, Nietzsche, Cohen, Benjamin bis Donna J. Haraway.* Vorwerk 8.

Döring, N. (2014). *Warum Sexting unter Jugendlichen (k)ein Problem ist. Online verfügbar unter.* https://www.medienbewusst.de/handy/20140729/warum-sexting-unter-jugendlichen-kein-problem-ist.html#more-36290, last viewed 25.02.2020.

Döring, N., & Mohseni, M. R. (2018). Male dominance and sexism on YouTube: results of three content analyses. *Feminist Media Studies, 19*(4), 512–524.

Eickelmann, J. (2017). *Hate Speech und Verletzbarkeit im digitalen Zeitalter: Phänomene mediatisierter Missachtung aus Perspektive der Gender Media Studies.* Transcript Verlag.

Foucault, M. (2000). Die Gouvernementalität. In U. Bröckling, S. Krasmann, & T. Lemke (Eds.), *Gouvernementalität der Gegenwart. Studien zur Ökonomisierung des Sozialen* (pp. 41–67). Suhrkamp.

Fröhlich, G., & Rehbein, B. (Eds.). (2008). *Bourdieu-Handbuch: Leben - Werk - Wirkung.* Metzler, J B.

Gesellschaft für Medienpädagogik und Kommunikationskultur in der Bundesrepublik Deutschland e. V. (GMK). (2013). *Medienkompetenzförderung für Kinder und Jugendliche*. http://medienkompetenzbericht.de/pdf/Medienkompetenzfoerderung_fuer_Kinder_und_Jugendliche.pdf

Hoffarth, B. (2009). *Performativität als medienpädagogische Perspektive. Wiederholung und Verschiebung von Macht und Widerstand*. Bielefeld: Transcript.

Iske, S. (2014). *Medienbildung*. https://content-select.com/de/portal/media/view/53bcf17b-0408-4cf2-937b-1bd22efc1343

Jane, E. A. (2017a). 'Dude ... stop the spread': antagonism, agonism, and #manspreading on social media. *International Journal of Cultural Studies, 20*(5), 459–475.

Jane, E. A. (2017b). Gendered Cyberhate: A New Digital Divide? In Massimo Ragnedda & Glenn W. Muschert (Eds.), *Theorizing Digital Divides* (pp. 158–198). Oxon: Routledge.

Jörissen, B. (2011). *Medialität und Subjektivation.: Strukturale Medienbildung unter besonderer Berücksichtigung einer Historischen Anthropologie des Subjekts*. Habilitationsschrift, Universität Magdeburg, Magdeburg.

Jörissen, B. (2014). Digitale Medialität. In C. Wulf & J. Zirfas (Eds.), *Handbuch Pädagogische Anthropologie* (pp. 503–513). Springer.

Kaspar, K., Gräßer, L., & Riffi, A. (2017). *Schriftenreihe zur digitalen Gesellschaft NRW: Band 4. Online Hate Speech: Perspektiven auf eine neue Form des Hasses*. kopaed VerlagsGmbH.

Kerres, M. (2018). *Mediendidaktik: Konzeption und Entwicklung digitaler Lernangebote*.

Koller, H.-C. (2011). *Bildung anders denken: Eine Einführung in die Theorie transformatorischer Bildungsprozesse. Pädagogik*. KOHLHAMMER.

Krämer, S. (2002). Sprache - Stimme - Schrift. In U. Wirth (Ed.), *Performanz. Zwischen Sprachphilosophie und Kulturwissenschaften* (pp. 323–346). Suhrkamp.

Laclau, E., Mouffe, C., & Hintz, M. (2012). *Hegemonie und radikale Demokratie: Zur Dekonstruktion des Marxismus. Passagen Philosophie*. Passagen Verlag.

Lemke, T. (2004). Test. In U. Bröckling, S. Krasmann, & T. Lemke (Eds.), *Glossar der Gegenwart* (1st ed., pp. 263–271). Suhrkamp.

Lemke, T. (2006). Die politische Theorie der Gouvernementalität: Michel Foucault. In A. Brodocz & G. S. Schaal (Eds.), *UTB für Wissenschaft: Uni-Taschenbücher: ... Politische Theorien der Gegenwart. Eine Einführung* (2nd ed.). Budrich.

Lumsden, K., & Morgan, H. (2017). Media framing of trolling and online abuse: Silencing strategies, symbolic violence, and victim blaming. *Feminist Media Studies, 17*(6), 926–940.

Mersch, D. (2006). *Mediale Paradoxa. Zum Verhältnis von Kunst und Medien.: Einleitung in eine negative Medienphilosophie*. http://www.sicetnon.org/index.php/sic/article/view/115/130

Mitgutsch, K. (2009). Medienkompetenz weitergedacht: Über die Responsivität medialer Erfahrungen. In R. Kubac, C. Rabl, & E. Sattler (Eds.), *Weitermachen? Einsätze theoretischer Erziehungswissenschaft* (pp. 194–207). Königshausen & Neumann.

Pieper, M., & Gutiérrez Rodríguez, E. (Eds.). (2003). *Gouvernementalität: Ein sozialwissenschaftliches Konzept in Anschluss an Foucault*. Campus-Verl.

Reckwitz, A. (2008). Subjekt/Identität. In S. Moebius & A. Reckwitz (Eds.), *Poststrukturalistische Sozialwissenschaften* (pp. 75–91). Suhrkamp.

Reh, S., & Ricken, N. (2012). Das Konzept der Adressierung: Zur Methodologie einer qualitativ-empirischen Erforschung von Subjektivation. In I. Miethe & H.-R. Müller (Eds.), *Qualitative Bildungsforschung und Bildungstheorie* (pp. 35–56). Budrich.

Ricken, N. (1999). *Subjektivität und Kontingenz: Markierungen im pädagogischen Diskurs. Epistemata. Reihe Philosophie: Bd. 249*. Königshausen & Neumann.

Schäfer, A. (2004). *Theodor W. Adorno: Ein pädagogisches Porträt*. Beltz.

Schäfer, A. (2019). Bildung und/als Subjektivierung: Annäherung an ein schwieriges Verhältnis. In N. Ricken, R. Casale, & C. Thompson (Eds.), *Schriftenreihe der DGfE-Kommission Bildungs- und Erziehungsphilosophie. Subjektivierung. Erziehungswissenschaftliche Theorieperspektiven* (pp. 119–136). Beltz Juventa.

Schäfer, A., & Thompson, C. (2010). Negative Dialektik. In T. W. Adorno (Ed.), *Schlüsselwerke der Identitätsforschung* (pp. 141–159).

Sesink, W. (2014). Eine kritische Bildungstheorie der Medien. In W. Marotzki & N. Meder (Eds.), *Medienbildung und Gesellschaft: Vol. 27. Perspektiven der Medienbildung* (pp. 11–44). Imprint: Springer VS.

Wischmann, A. (2018). The absence of 'race' in German discourses on Bildung. Rethinking Bildung with critical race theory. *Race Ethnicity and Education, 21*(4), 471–485.

Zima, P. V. (2000). *Theorie des Subjekts: Subjektivität und Identität zwischen Moderne und Postmoderne*. Francke.

CHAPTER 5

Powerful Entanglements: Interrelationships Between Platform Architectures and Young People's Performance of Self in Social Media

Viktoria Flasche

Abstract This chapter explores intertwinements between digital media and communicative and socio-cultural practices as they emerge in relation to contemporary cultures, specifically youth cultures. Social media platforms such as Facebook, Instagram and TikTok are discursive-operative networks within a framework of economic strategies. The chapter's empirical approach draws on the assumption that young people's aesthetic practices, transmitted via social media formats, evoke in each instance specific relational modes that preform a space of possible subject positions. The chapter summarises the findings of two selective longitudinal studies examining young people's practices of self-articulation, consistently interpreted in the context of the specific platform used in each instance. These findings point to the potential of aesthetic-tentative practices as performed by young people to catalyse societal critique.

V. Flasche (✉)
Institut für Pädagogik, FAU Erlangen-Nürnberg, Erlangen, Germany
e-mail: viktoria.flasche@fau.de

P. Bettinger (ed.), *Educational Perspectives on Mediality and Subjectivation*, Palgrave Studies in Educational Media,
https://doi.org/10.1007/978-3-030-84343-4_5

Keywords Youth cultures • Social media • Media education • Subjectivation • Reconstructive social research

1 Introduction

Social networks provide a range of possibilities for individuals to express themselves within a personalisable virtual space; in so doing, incidentally to their primary economic purpose, they offer us role models and 'tribes' with which we may align ourselves socially. The current generation of young people[1] is the first to have grown up with social media networks as a ubiquitous and everyday part of life (Röll, 2014, Lovink, 2012: 37; MPFS, 2018). Negotiating friendships has always been a part of teen life. Prior to the social media age, however, most of these negotiations took place implicitly, without explicit labels (Boyd, 2008). Unlike adults, faced with relearning aspects of their public behaviour due to the impact of networked technologies, teens simply learn from the outset how to handle themselves publicly with networked audiences in mind (ibid., p. 295). Young people's contemporary lifeworlds are in effect post-digital—to note this is not to imply that digitalisation is already complete, but rather to acknowledge that, in light of the complex digital transformation of all aspects of everyday life and of its diverse manifestations, the analogue/digital distinction is no longer enough (Cramer, 2014). In addition to this, the digital documentation in social media of practices which face-to-face interaction cannot pin down opens up a research opportunity for empirical approaches (Manovich, 2012; Schreiber & Kramer, 2016).

This article explores the intertwinements between digital media and communicative and socio-cultural practices as they emerge in relation to contemporary cultures, specifically youth cultures (Joerissen, 2018; Hugger, 2013). Its particular emphasis, from the point of view of theories of *Bildung*[2] (Joerissen, 2018; Bublitz, 2014; Richter & Allert, 2017), is on the ways in which social media practices generate identities and subjectivities. My central interest is in identifying which processes of subjectivation unfold in the field of tension between the socio-technical structure of the social media platform in question and the individual's relationship with the self as reflected in multimedia representations of that self. The work underlying this chapter was a selective longitudinal examination of aesthetic practices of self-representation as engaged in by young people between 2013 and 2019, taking into account the socio-technical architectures of the platforms they used. The analysis that follows encompasses two empirical projects, a study of Facebook profiles (Flasche, 2017a, b, 2018) and the qualitative findings of the (Post-) Digital Cultural Youth Worlds (DiKuJu) study undertaken in Germany between 2016 and 2019.[3] The extended analysis also includes older studies which explored such practices.

2 Context and Theoretical Viewpoint: Subjectivation in the Field of Tension Between the Medium and Its Practices

Social media platforms are part of dynamic networks with flexible structures; they form the new morphology of the current social order. We might operationalise the concept of the dispositive (Foucault, 2008, p. 199) to the end of perceiving them as networks of conditions, "expansions and sediments of power" (Reichert, 2008, p. 14).[4] Media dispositives are subject to historical cycles, and media historiography can enable us to analytically access processes of subjectivation (ibid.). Compared to the discourse concept, the concept of the dispositive expands the perspective: "While—metaphorically speaking—the discourse analyst wants to 'discover' the conditions and rules of the practice of making statements through and beyond the statements, and from there draw conclusions about the consequences of the 'true knowledge' processed in this way, from the perspective of dispositive theory the statement formations in their spatio-temporal situation form the analytical starting point of the research perspective" (Schneider, 2015, p. 31).

Any analysis of contemporary aesthetic practices communicated via media will need to take account of their historicity, focusing on points or phases of transition rather than on supposedly absolute moments of rupture and upheaval: "The performative and subversive energy of social practices of appropriation [and adoption of new forms of media] makes for an inhomogeneous, dystopic and divergent tectonic landscape when it comes to technical ruptures in media [formats and use]" (ibid., p. 18). Since the turn of the last millennium, "protocol-logical networks" (Galloway & Thacker, 2007) have become the new social normality. Their databases are spaces for the doubtless hegemonic construction of categorical identities and collectives, with the effect of pre-structuring modes of perception and practice (Jörissen, 2020, p. 351).

Social media platforms such as Facebook, Instagram and TikTok are discursive-operative networks within a framework of economic strategies. They create reactive media spaces of the convergence of various divergent forms of media, such as text and still or moving images. This means that those operating the platforms, who are active in economic intent and interest, create reactive and responsive virtual-creative space by means of processes largely controlled by algorithms. Reactivity, in this sense, refers to the constitutive dependence of social media platforms and of their algorithmic decision-making processes on the activities of their users. Pointing to the findings of foundational information research, Nassehi certifies that the algorithms currently used in some social media are capable of developing "selective intrinsic activity" (Nassehi, 2019, p. 238). These algorithms are able, on the basis of user activities, to abductively draw their own conclusions that go beyond what human actors have programmatically designed (ibid., p. 241). These deep learning processes have reinforced the role of software and hardware as co-constructors of images and practices of self-presentation among young people (Schreiber, 2017). Schreiber was able to demonstrate empirically how smartphones and specific social media apps influence the images young people produce. An ability of these apps to increase user activity via reactive algorithmic decision-making processes would augment this influence still further. One already evident upshot of this is that platforms such as TikTok, for example, can place advertising in the stream of images in such a way that it is almost impossible to identify it as such.

The platforms analysed in the research set out here all work on the basis of the same business model, which primarily pursues profitable operation via increasing the number of people staying and engaging on the platform in order to target them with personalised advertising. A further aspect of this business model relates to selling on the data collected via the platform, largely for market research purposes. The platforms' interfaces act as hubs for these processes and constitute the designed, formalised framework that defines the possibility and impossibility of interactions. In investigating their activity in this context, we need to perceive them as a product of all their informational and structural features; this will enable us to identify how complex social mechanisms such as the establishment of a shared or ubiquitous practice materialise—or dematerialise—in algorithmically controlled operations.

The focus on operationalisation and monetisation which characterises these platforms calls for the addition of a subjectivation-related perspective to a theory-based analysis of media practices: Each platform generates specific scenarios of invocation to the user's self. Although the self is susceptible to the appropriating and controlling action of various techniques imposed from without, it is also capable of evading institutional access and inventing new forms, affects and intensities of being.

3 A Trans-Actional Research Perspective

The evocative character of media is a central factor in the processes of subjectivation described by Bublitz (2010) following Butler (2003): "Media are loci both of the individual's repeated linguistic and visual self-presentation and self-representation and of the subject's self-production in both senses of the term—a self-presentation and self-representation as a social subject." (Bublitz, 2014 p. 12). In this context, I have based my empirical approach on the assumption that the aesthetic practices of young people, transmitted via social media formats, evoke in each case specific relational modes that preform the space of possible subject positions. The relationally integrated practices of a young person serve as a case in which we observe this evocation: "The prototypical character immanent to things, the structurally immanent knowledge they embody, brings its influence to bear both on the 'possibilities of their use' and, at a theoretically more profound level, their relational potential with regard to other (material or human) actors. In the context of everyday practices, things, and environments of things, represent invitations to subjectivation, specifically, therefore, invitations to become user-subjects (consumers, audiences, prosumers etc.) in a particular way." (Joerissen, 2015, p. 218). Following this view, we may identify the trans-actional nature of young people's media-based practices, their location in the space between the practice and its medium. In this sense, my qualitative approach aims at the reconstructive analysis of this in-between and draws on the analysis both of the specific platform's socio-technical architecture and on the articulations made within this by young users.

Orders of visibility on social media applications are subject to the '*creativity dispositive*' (*Kreativitätsdispositiv*) (Reckwitz, 2015), which enables specific self-positionings of the subject. Reckwitz' analyses in this context refer to changes in Western societies since the end of the twentieth century (ibid., p. 1). The characteristic feature of the previously dominant 'rationalisation dispositive' was its orientation towards schematising and standardising subjects and things. The creativity dispositive, by contrast, centres on audiences, singularities and intensified affectivity: "This dispositive, at its core, is a configuration of the social in which subjects gather as an audience for objects or other subjects which make a significant or sensory impression and simultaneously render themselves performers [or potential performers] to such an audience. Neither objects nor subjects are formed here as replications and repetitions of the same, as in the rationalisation dispositive, but instead [emerge] as singularities, i.e. as non-comparable instances of the particular". (ibid., p. 2).

In this way, viewed from a Foucauldian perspective emphasising power, it is evident that all social media platforms act as "subjectivation apparatuses" leading "to the voluntary self-exploitation of private life" (Wueschner, 2019, p. 254). However, this view runs the risk of losing sight of productive and transformative moments in the practices associated with social media, which their users encounter with an aesthetic attitude that fundamentally follows a playful premise "and makes the end in itself of simulation strong against the purpose as a means of communication" (ibid., p. 256).

4 Methodological Approach and State of Research: The Recent Development of Young People's Practices of Self-Presentation Practices

This chapter seeks to illuminate its subject via analysis of findings obtained from two distinct studies. The first took place on a corpus of 428 Facebook profile photographs of young people collected in mid-2013 from Facebook groups formed under the names of German secondary schools. The central interest was in identifying which processes of subjectivation unfold in the field of tension between the socio-technical structure of the social media platform in question—in this case, Facebook—and the individual's relationship with the self as reflected in multimedia representations of that self. In line with the heuristics developed in the context of the study's research question, the entire corpus has been encoded. Classification of the corpus took place in accordance with the 'serial-iconographic' photograph analysis method described by Pilarzcyk and Mietzner (2005). The core of this approach is the reciprocal relationship of two procedures to each other, these being the iconographic/iconological interpretation of individual images—following Panowsky (1983) and Imdahl (1980)—as well as the serial analysis of entire collections of photographs. The aim of this combination is to test and quantify the hypotheses obtained in the detailed analysis on a larger corpus of images. Pilarczyk and Mietzner thus aim in their method for a *via media* between qualitative and quantitative research logic (ibid., p. 131).

Other changes have taken place in terms of social media use by age group. In the 1990s, Turkle (1995) was able to describe multi-user-media applications as a moratorium, as a free space for virtually trying out alternate roles. As late as 2008, Boyd stated that most of the young people she had interviewed conceived of MySpace and Facebook as "effectively teen space" (Boyd, 2008, p. 290). Seen from today's vantage point, these views, suggesting as they did that social media had the capacity to undermine adult control over young people's lives, conceptualise the *old* internet, shaped by liberating and participatory values (Lovink, 2012). The economically-driven properties of today's social media platforms renders these ideas outdated. Adults have taken to—some might say taken over—the platforms previously used by teenagers, which, in line with their reactive design, have partially responded to this takeover. To a degree, however, new platforms have emerged that specifically address teenage needs for a space in which adult control is suspended. One example is Snapchat, whose response to young people's experience of consistent visibility online has drawn precisely that constituency to it (MPFS, 2018).[5]

It is evident, then, that social media practices and applications have diversified to a striking extent between 2012 and the present moment. Accordingly, the first study in the context of my doctoral research, whose findings this chapter discusses, bears of necessity the caveat that the cultural patterns it has captured may already have disappeared when these findings see the light of day. The speed of change in this field effectively consigns detailed case studies to historiography (Lovink, 2012, p. 15). The German "Shell Study" for 2019 also states that constant change in media use is typical of the current generation of young people, and that studies in this area very quickly become outdated (Albert et al., 2019, p. 40).

It is for this reason that this detailed discussion of this study's findings have to take into account those of a second study, "Postdigital Cultural Youth Worlds", which ran between 2016 and 2019, investigating the social impact of the digital transformation on young people's current artistic/creative practices, their cultural education and their engagement in cultural life and activities. In parallel to a quantitative representative interview study, the research design encompassed qualitative online surveys, expert interviews and group discussions with young people, casuistic analyses of selected individual cases, and a methodological outline of the specific OpenSpace 'Barcamp' format.[6] The project's first, exploratory phase concentrated on developing an overview of young people's (post-)digital aesthetic practices, primarily employing descriptive methodological approaches. The work included expert interviews with professionals from the field of cultural education and group discussions with young people recruited from culture-related institutions (schools with a cultural focus, young people's media centres). The findings of this research served as a basis for the design of the items in the quantitative sub-project, alongside enabling further targeted surveys conducted as theoretical sampling and providing the project outcomes with greater nuance and depth. Reconstructive in-depth analysis took place on young people's cultural practices emerging in the context of the Barcamp format and transactional interviews held during a digital festival (Joerissen et al., 2020). Thus far, education studies have primarily operationalised transactional perspectives for the purpose of incorporating spaces and material worlds of things in research (Nohl, 2017). This transactional research methodology opens up an empirical view of the constitutive property of physical, material and spatial actors in its capacity "to reconstruct the genesis of these entities from transactional practices" (ibid., p. 1). This approach to research centres reconstructive artefact analysis and, put in terms of actor-network theory (Latour, 2006), engages with both human and non-human actors on an equal level. With regard to the research field of the two studies described here, an analysis of social media platforms takes the place of artefact analysis. The use of this approach, in a modified form, is restricted to where the analysis of hardware, such as smartphones and the corresponded software. The findings detailed in this chapter will consist primarily of these transactional interviews.

Together, these two distinct pieces of research provide a selective longitudinal section of the practices by which young people articulate their ideas of their selves, consistently interpreted in the context of the specific platform used in each case. The discussion that follows will supply an overview of the most significant aspects of the analyses and combine them with central findings from the analysis of still and moving images and from interviews. The transactional methodology employed here thus enables the illumination of entanglements between social media platforms and their users' practices.

The examination of the findings employed the dimensions of social media analysis proposed by van Dijck and Poell (2013), which distinguish between the levels of the user, the business model, content, ownership, governance, and technology, the latter sub-divided in turn into (meta)data algorithms, protocols, interfaces and defaults. Interfaces had a particular role in the analysis. An interface acts to bring together formalised frameworks for possible interactions, i.e. it is the product of all informational and structural features of the platform and a pre-formalisation of the practices that take place on it. At the interfaces of the practices, only those practices are possible for which the settings provide. Even if algorithmic decision-making processes are able to restructure themselves reactively with regard to user activity, all possible activities are only possible in categories created in line with the platform's logic. The possibilities for interaction designed within the interface elicit processes of subjectivation. This space entails the design of specific user-subjects required to behave in a certain way if they are to 'act with the app'; one example might be following the app-imposed compulsion to present/display/represent themselves visually.

5 Excursus: Socio-Technical Development During the Period Under Study Using Facebook as an Example

Evaluation of the classifications obtained needs to proceed in consideration of Facebook's specificity as a social media platform. The social media age commenced with market leadership held by specific distinct platforms, such as Facebook, MySpace and (in the German-language context) the student platform StudiVZ, especially among teenagers and young adults. At the present time, by contrast, young people's social media activity typically takes in a number of networks and platforms (MPFS, 2020, p. 31). Until about 2015, Facebook was the social network with the largest number of members, which it associated with a claim to represent the present and future of discourse on social issues (Miller, 2012, p. 10). Miller's analysis of Facebook usage—albeit stemming from a strongly location-bound ethnographic analysis—formulates optimistic theses suggesting the capacity of social media to drive positive societal transformations. Alongside this optimism, a view arose which emphasised the market-shaped channelling of communication on Facebook in particular. In 2012, one year before the study we refer to above, Facebook's rigid and biographically-based appeal to the self (Wiedemann, 2011) continued to dominate the social media landscape. According to Lovink (2011, p. 183), the platform was instrumental in establishing a "culture of self-revelation" and of the management of the self as a central mode of social media use. In the course of 2014, however, it lost its market leadership for young Americans aged between 13 and 19 (Piper Jaffray Survey, 2014); currently, Facebook is ranked only the eighth most popular social media site among 12- to 19-year-old Germans (MPFS, 2018, p. 35). At the time of my first study Facebook held a form of monopoly position in young people's social media use and served to network and manage various sub-systems, such as Instagram. As a company, Facebook formally owns current market leaders such as Whatsapp and Instagram; as a platform, it has, as indicated above, lost its supremacy as a social medium and as the priority tool for managing 'collapsed contexts' (Boyd, 2014, p. 31). In this sense, on the specific level of social media interaction, Facebook has retreated backstage, controlling the current 'front stages' in terms of IT and organisation.

During Facebook's golden age, although around three-quarters of 12- to 25-year-olds were aware of the danger associated with the publication of personal data online and large businesses, insofar as their policies became public, tended to view Facebook at least critically, the platform's use nevertheless appeared to them as a 'must-do' (Albert et al., 2015, p. 130). The numerous interviews Boyd conducted with young people appeared to uncover a key reason for this, namely that Facebook served its young users as a "context manager" via which they handled their online relationships with family, school and various circles of friends; one result was that these young people's profiles tended to be completely public to broad groups of other users (Boyd, 2014, p. 32). At an explicit level, the front stage, Facebook attempted to market itself as a platform for "real" people (Zuckerberg, cited in Boyd, 2014, p. 50). At the implicit level, however, it never seriously tried to realise this claim. Although it asked users to use their real names and deleted any fake pages brought to its attention, it was always possible for users to run several pages at the same time with little effort and thus to operate decidedly selective identity management. Facebook has now made changes to its security settings; where they previously inherently implied a rigid regime of visibility, they now offer a wider range of options than in 2012. For example, it has become easier for users to personalise their security settings and thus control the range of their visibility on the platform; there have also been changes to default security and visibility settings. It is important, however, to note the commercial impetus behind this development; technical advances have made it possible to process large volumes of metadata at great speed and therefore place 'personalised' advertising even without strict profile management. Facebook no longer writes to all users with obviously made-up names and threatens to delete their profiles if they do not switch to using their real names. New users creating their profiles are no longer exhorted to show *themselves* as Wiedemann (2011) has worked out from her analyses. Even without the use of real names or images, tracking of past activities, contacts and other details can be used to derive data of relevance to advertising.

As noted above, some of the platforms currently at the top of the popularity tree, such as WhatsApp and Instagram, are part of the Facebook consortium, minimising any impact of the loss of importance sustained by the Facebook platform between 2014 and 2016. We might, then, reconstruct from this the assertion that Facebook was able to release its hold on the front stage because it has doubly secured its position of backstage power.

6 Findings: The Permanent Progression of Platforms from Static Profiles to Diversified Stream Portfolio

In order to illustrate the key insights emerging from the before mentioned studies, I will compare two platforms: Facebook, and the currently highly popular „TikTok" (on this platform's success among young people in particular, cf. Reuter & Koever, 2019). As outlined in the excursus, Facebook has lost its monopoly position and its use by young people in particular is limited; the aesthetic pattern of articulation the research reconstructed in its case is also in evidence on other platforms. I will pay specific attention here to the patterns of 'bricolage' and of the 'mask'. Since 2010, social media sites such as Instagram and Pinterest have responded to the increase in posting of images online by centring their entire infrastructure on images, to which text is compelled to relate. A user's profile picture plays only a subordinate role; in terms both of the platform's structure and of the interaction that takes place on it, a multiplicity of other pictures comes to dominate. These platforms diversify the possible range of communication via images by using image clusters and temporary, auto-deleting compilations of images ('stories'). Platforms such as Snapchat effectively go further by, first, permitting only the temporary showing of images, and second, facilitating the use of filters in such a way as to foreground the playful, carnivalesque mode (Levin, 2015). The first study discussed here focused its attention on Facebook profile pictures because of the special role they held within the platform architecture at that time. In contrast to Turkle's hope for the internet as a moratorium, Facebook has been, at that stage in its existence, not a network which permitted the cultivation of anonymous or fictional alter-egos or, arising from this, the testing out of different identities in a playful way (Boyd, 2014, p. 41). Other networks, such as those that accompany certain computer games, provided spaces for this type of usage as did Twitter (cf. Boyd, 2014, p. 204). On Facebook, the user's profile picture appeared as a frontispiece, as it often remained the first and for outsiders the only impression, and thus acted as a decodable condensation of world and self-perceptions (Flasche, 2017b, p. 272). Rigid invocations accompanied the upload process; within the entire corpus, only 7 percent of the users failed to post an image.

Structurally present neither in the media structure of the Facebook profile image nor in the platform's rigid identity politics, the carnivalesque mode nevertheless occurred, in clusters, within both of these entities, which points to the productive—in some cases subversively articulated—character of the profile image's empty space. The first study showed that young people in particular used bricolage (Flasche, 2017a) and masks (Flasche, 2020) to take a performatively intractable position against the order of visibility pre-formed in the platform. Counteracting the stage preset for them by the Facebook profile by setting backstage practices,[7] they used external image processing programs to split their profile picture into individual images and to put several identities into one (Fig. 5.1).

In practising casual pictorial techniques, they refused the interpellations of Facebook's default settings, also reconstructed as governmental, which demanded they *show themselves in the best possible light* (Wiedemann, 2011; Flasche, 2018). The quantified analysis of profile pictures, conducted as part of the serial iconographic method referenced above (Flasche, 2017b), showed that these young people presented themselves primarily as *networks* of things, spaces and other actors, reminding of Latour's Actor-Network-Theory (1998). A key element of the self-representations in

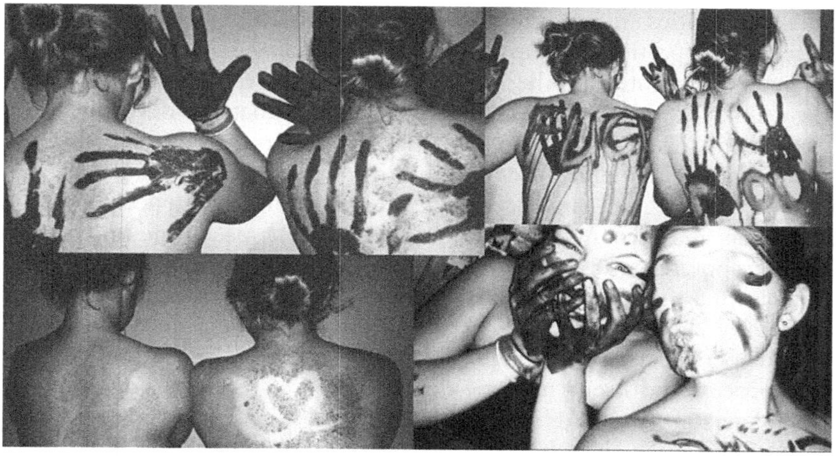

Fig. 5.1 Sample image from the first study: Facebook profile photo (anonymised) © Viktoria Flasche

evidence in the images is the action-related dimension of connectivity in a twofold sense: first, on the level of the image itself, and, second, on the level of the platform on which the user posts the image so it might be immediately networked, linked and affixed with attributes such as hashtags as provided for in the platform's architecture (Flasche, 2017b; Schreiber & Kramer, 2016).

This analysis points to the practices that are currently dominant on TikTok, where soundtracks, miniature-dialogues, film clips and music sequences can be linked to a video usually recorded by users themselves. The app is the central medium for linking, saving, sharing and commenting on these videos. The typical view of this platform is a tiled view of a vertically running video web feed. The user sees a constantly moving and growing sequence of small vertically aligned rectangles. The very short videos (15 seconds maximum) published by the users appear in a display individually controlled via algorithms.[8] As soon as one as created a profile, one will only see an individually created stream based on the previous personal activities. There is thus no "neutral" view that is equally available to all users. The display of the app's browser version, viewed without being logged into a profile, is personalised on the basis of the user's IP address. These "mini-mini-videos" establish a "temporary undertow by which further activities are set in motion" (Porombka, 2016, p. 32). The videos enter into a particular relationship with affect and sensory corporeality: "They concentrate on [the representation of] faces, postures or movements that express anger, sadness, despair, resignation, devotion, love, excitement, disgust [...]. Because they do this in a loop, repeating the same expression and gesture over and over again, they have a particularly intense effect [...]" (ibid.). This leads to a flow experience, in which awareness and certainty about one's own, present space-time-body structure evaporates and gives way to an affect-guided, almost magical "involvedness" (ibid.; Carnap & Flasche, 2020). The practices in action on TikTok show how the TikTok algorithm, which suggests music titles in various personalised categories, is centrally inscribed in the sound selection process and users' mimetic desire. The user receives suggestions in several personalised feeds such as the 'For You' feed. The videos are placed algorithmically in an order that aims to generate increased length of stay and activity. This process usually remains hidden from users, that is, they receive no explicit information that it is taking place. This becomes particularly evident in the interviews, where references to popularity of content fail to differentiate between what the interviewees themselves *like* and what has received many "likes" from other users. Like an *amoeba* (ibid.), the algorithm only reveals itself to perception in changing manifestations. The metaphor of the amoeba illuminates the non-transparency and ungraspability of the algorithm's operation to users; its limits, modes of operation and actions elude comprehension or description. This inconceivability obscures from gaining any knowledge as to their positioning in relation to the algorithm or indeed as to how they are positioning themselves. The transactional interviews showed that when users describe their creative processes, the boundaries between decisions taken by the algorithm and their actions remain unclear to them. The amoeba-like algorithm, as a changing mutant entity, constantly changing its form, protrudes into their practices, without them being able to describe it as such, because it constantly escapes their perception.

Alongside this, the DiKuJu study identified practices that significantly exceed the algorithmic decision-making process by recombining different genres such as videos, music snippets, comedy dialogues and games as well as different platforms, each with their specific structural logic. One example of this consists in videos co-created by young people together with their "online best friends" in the multiplayer game Minecraft, recording play by means of screen recording or with their smartphones. The formal

Fig. 5.2 Example from the second study: Compilation of screenshots from a TikTok video. © DiKuJu project

design of these videos contrasts with typical TikTok videos: The cuts are fast and fragile, the videos flicker and change abruptly in brightness and colour (ibid.). In picture quality, as in their plot and music, they reject the viewer's gaze and expectations almost aggressively. Transgression does not occur here as a process reflected on, due to the amoeba's elusion of comprehension, but instead as a collaborative interplay (see Fig. 5.2).

These practices on TikTok transcend the logics of individual platforms in an exuberant, explosive structure of forms and allow antagonistic gestures that transcend mere (innocent) "involvedness" (Porombka, 2016) and the "feelgood atmosphere" typical of TikTok (Reuter & Koever, 2019).

7 Conclusion: Transformative Theory of *Bildung*—And Powerful Entanglements

This chapter ends with the open question of how, with regard to the studies' findings, we might comprehensively rethink media education in the light of their implications. If pedagogical endeavour seeks to create reflective distance or unfold critical and thus possibly creative potential, the insights into current media practices garnered by the studies pose the question as to how this might succeed when the object of such prospective critique, that is, a distinct media format, is so elusive. Platforms and their constituent elements are no longer an object in an epistemological sound sense, but rather an entity that has always already been interwoven with the subject. In this sense, only a heuristic understanding can speak of media *use* here. This use of language, and of academic semantics, has left the terrain of neutrality and suggests a reading of media practices that cannot incorporate current developments such as so called "deep-learning"-algorithms (Jörissen, 2020, p. 348).

The selective longitudinal overview described in this chapter offers us a number of insights towards a response to this open question that does not, however, reach to a final conclusion. Despite their limited scope, the cases described provide indications that critical—although not necessarily reflective—potential arises where the platforms' ‚protocol logics' becomes visible to us, as explicit in the articulatory media practices of young people. This often happens in instances which confuse or 'throw' the viewer, disrupting their viewing habits. These videos are often difficult to watch, they flicker, they cut abruptly, their sequences are too short, they appear illogical to us by seemingly randomly combining sounds and images. This intractability arises as a function of our habitual and above all generational viewing habits. From an empirical perspective it is precisely in this moment of the viewer's being 'thrown' that the potential arises to bring the visual and structural logics of the platforms forth out of their invisibility.[9] We might classify such practices as critical in an *aesthetic-tentative* manner rather than in a *rational-reflective* form. Their potential for *Bildung* lies precisely in "expanding [our] scope for action and experience" (Richter & Allert, 2017, p. 251) and in not resolving conflicting positions, but keeping them open. Thus, the results of the longitudinal section do not point to the productivity of models of 'digital literacy' as presented in administrative educational contexts (Jörissen, 2020, p. 348). The reference of the empirical findings to previously formulated theoretical approaches suggests the utility of increased attention to what we might call the creative/tentative game in the context of current media practices, alongside the practice of reflective, critical distancing. Together, these ways of exploring invisible and flowing, constantly evolving protocol logics may show a way forward in education on this specific and crucial area of modern-day media. The consequence of this would be that media-pedagogical action would have to aim at opening up contexts in which deconstructive aesthetic strategies could be tested and in the best case even validated.

Concluding, I note that existing theoretical work towards a transformational theory of *Bildung* remains important and may also help guide us along this path. Richter and Allert (2017, p. 252) describe the range of dynamic educational processes in a digital culture as “poetic plays” that lead to productive entanglements. Koller attempts not only to think differently about the educational process, but also to refigure the educational process itself as “thinking differently” or “becoming different” (Koller, 2018, p. 9). In the theory of structural media education (Joerissen & Marotzi, 2009), this openness is built-in integrally where tentativity in particular appears as a decisive moment in educational processes: “Uncertainties must be given a place, better, several places in our thinking; then, and only then, will a tentative, experimental, replaying, testing, innovative, category-inventing, creative processing of experience become possible” (ibid., p. 21).

Notes

1. The young people that are the subject of the work detailed in this chapter were aged 12–24 in the period of investigation, from 2013 to 2019, and therefore on the boundary between Generations Y and Z. Generation Y comprises the cohorts born between 1985 and 2000, whom a Deutsche Shell study (2002) described as the “pragmatic generation” in an age marked by uncertainty and upheaval. This generation, too, was centrally shaped by Rapid advances in information and communication technologies brought a decisive influence to bear on this generation; but it is only for those born after 2000, termed Generation Z, that the digitalisation of broad areas of life has been all-determining (Albert et al., 2019). Generation Z, also described as “iGen” (Twenge, 2017), is the first successor generation of the “digital natives”, for whom smart technologies and social media are ubiquitous features of daily life in a process of permanent change (Albert et al., 2019, p. 40).
2. ‘Education’ is an at best approximate rendering of the German concept of *Bildung* (Horlacher, 2016).
3. Full title: (Post-) Digital Cultural Youth Worlds—Development of New Methodological Instruments for the Development of Research on Arts and Cultural Education in the Digital and Post-Digital World/DiKuJu, research project funded by the German Ministry of Education and Research, 2016–2019, led by Prof Dr. B. Jörissen and Prof Dr. S. Keuchel (Joerissen et al., 2020).
4. All textbook translations by the author.

5. There are additionally micro-generational distinctions alongside the broader adult/non-adult divide. Analysis of group discussions from the DiKuJu study illuminates this phenomenon, with generational stratifications evident within the groups: the older participants (16–20) express surprise and concern at the behaviour of younger ones (12–16) in relation to Snapchat and there is no shared framework context as regards media and app use across the groups' age range of 12 to 25.
6. A "Barcamp" is originally a conference method from the field of technology research. As a so-called *un*conference, the BarCamp has no predefined talks or workshop topics, but is rather designed in a collaborative manner by the participants themselves. For the survey in the context of the DiKuJu project, a three-day BarCamp on the topic of "Who are we?", attended by 46 young people, was held at the Academy for Cultural Education of the German Federal Government and the State of North Rhine-Westphalia in Remscheid. (Joerissen et al., 2020).
7. 'Backstage practices' are those activities that are not likely to be carried into *public* space or the *front* stage. In the context of these images, they are motifs which do not 'stage' or 'present' the user, such as blurred snapshots taken in front of a bathroom mirror (Flasche, 2018).
8. The default settings allow only videos with a maximum length of 15 seconds. These presets can be fine-tuned to allow up to 60 seconds. However, the majority of videos are aligned with the 15 seconds.
9. A similar effect can be observed where users try to subversively place censored political messages on TikTok. A famous example, which can now be found on YouTube, would be Feroza Aziz's critique of Chinese policy towards the Uighurs, which she recorded in the style of a make-up video (https://www.youtube.com/watch?v=EcqfCgCCLDk; last accessed 12.07.2020).

References

Albert, M., Hurrelmann, K., & Quenzel, G. (2015). *17. Shell Jugendstudie.* Frankfurt/M.

Albert, M., Hurrelmann, K., & Quenzel, G. (2019). *18. Shell Jugendstudie. Eine Generation meldet sich zu Wort.* Basel.

Boyd, D. (2008). *Taken out of context.* www.zephoria.org/thoughts/bestof.html

Boyd, D. (2014). *It's complicated. The social Lives of networked teens.* New Haven/London.

Bublitz, H. (2010). *Im Beichtstuhl der Medien.* Bielefeld.

Bublitz, H. (2014). Im Beichtstuhl der Medien. Konstitution des Subjektes im oeffentlichen Bekenntnis. *Oesterreichische Zeitschrift für Soziologie, 13*(38), 7–22.

Butler, J. (2003). *Giving an account of oneself.* New York.

Carnap, A., & Flasche, V. (2020). Diskursive Sichtbarkeiten—Auffuehrungen von Geschlechtlichkeit in (post)digitalen Jugendkulturen. In B. Hoffahrt (Ed.), *Feministische Perspektiven auf Medien*. Frankfurt/M.

Cramer, F. (2014). What is 'Post-digital'? *APRJA, 3*(1), 10–25.

Deutsche Shell. (2002). *Jugend 2002*. Frankfurt/M.

Flasche, V. (2017a). Jugendliche Bricolagen—Eine Spurensuche zwischen digitalen und analogen Raeumen. In M. Pietraß, J. Fromme, P. Grell, & T. Hug (Eds.), *Jahrbuch der Medienpaedagogik 14: Der digitale Raum, pp. 35–54*. Wiesbaden.

Flasche, V. (2017b). Das glaeserne Jugendzimmer? In I. Herrmann & S. Schinkel (Eds.), *Aesthetiken in Kindheit und Jugend. Sozialisation im Spannungsfeld von Kreativitaet, Konsum und Distinktion* (pp. 269–286). Bielefeld.

Flasche, V. (2018). Linkische Cyborgs—Jugendliche Selbstkonstitutionen als hybride Netzwerke. In A. Tervooren & R. Kreitz (Eds.), *Dinge und Raum in der qualitativen Bildungs- und Biographieforschung* (pp. 157–174). Opladen.

Flasche, V. (2020). Hinter den Spiegeln—Ikonische Selbstthematisierungen im Netz. In J. Fromme, S. Iske, D. Verständig, & K. Wilde (Eds.), *Big Data, Datafizierung und digitale Artefakte*. Wiesbaden.

Foucault, M. (2008). *Dispositive der Macht. Über Sexualität, Wissen und Wahrheit*. Berlin.

Galloway, A. R., & Thacker, E. (2007). *The Exploit: A Theory of Networks*. University of Minnesota Press.

Horlacher, R. (2016). *The educated subject and the German concept of bildung: A comparative cultural history*. New York.

Hugger, K.-U. (Ed.). (2013). *Digitale Jugendkulturen*. Wiesbaden.

Imdahl, M. (1980). *Giotto: Arenafresken: Ikonographie, Ikonologie, Ikonik*. Munich.

Joerissen, B. (2015). Bildung der Dinge, Design und Subjektivation. In B. Joerissen & T. Meyer (Eds.), *Subjekt Medium Bildung* (pp. 215–233). Wiesbaden.

Joerissen, B. (2018). Subjektivation und aesthetische Bildung in der post-digitalen Kultur. *Vierteljahresschrift für wissenschaftliche Paedagogik, 94*, 51–70.

Joerissen, B., & Marotzi, W. (2009). *Medienbildung—Eine Einfuehrung*. Stuttgart.

Joerissen, B., Schroeder, K., & Carnap, A. (2020). Postdigitale Jugendkultur: Kernergebnisse einer qualitativen Studie zu Transformationen aesthetischer und kuenstlerischer Praktiken. In A. Scheunpflug (Ed.), *Forschung zur kulturellen Bildung*. Wiesbaden.

Jörissen, B. (2020). Ästhetische Bildung im Regime des Komputablen. *Zeitschrift für Pädagogik, 66*(2), 341–355.

Koller, H.-C. (2018). *Bildung anders denken. Einfuehrung in die Theorie transformatorischer Bildungsprozesse* (2nd ed.). Stuttgart.

Latour, B. (1998). Ueber technische Vermittlung. In W. Rammert (Ed.), *Technik und Sozialtheorie* (pp. 29–81). Frankfurt/M.

Latour, B. (2006). Über technische Vermittlung. In A. Belliger & D. J. Krieger (Eds.), *ANThology*. Transkript.
Levin, A. (2015). Das vernetzte Selbst. In A. Bieber (Ed.), *Ego Update* (pp. 96–133). Düsseldorf.
Lovink, G. (2011). Anonymitaet und die Krise des multiplen Selbst. In O. Leistert & T. Roehle (Eds.), *Generation Facebook* (pp. 240–249). Bielefeld.
Lovink, G. (2012). *Das halbwegs Soziale*. Bielefeld.
Manovich, L. (2012). Trending: The promises and the challenges of big social data. In M. K. Gold (Ed.), *Debates in the digital humanities* (pp. 460–475). Minneapolis.
Miller, D. (2012). *Das wilde Netzwerk*. Berlin.
MPFS, Medienpaedagogischer Forschungsverbund Suedwest. (2018). *JIM-Studie 2017, Jugend Information Medien*. https://www.mpfs.de/fileadmin/files/Studien/JIM/2017/Studie/JIM2017_Gesamt.pdf
MPFS, Medienpaedagogischer Forschungsverbund Suedwest. (2020). *JIM-Studie 2019, Jugend Information Medien*. https://www.mpfs.de/fileadmin/files/Studien/JIM/2019/JIM_2019.pdf
Nassehi, A. (2019). *Muster. Theorie der digitalen Gesellschaft*. Munich.
Nohl, A.-M. (2017). Die empirische Rekonstruktion materieller Artefakte mit der Dokumentarischen Methode. In A. Tervooren & R. Kreitz (Eds.), *Dinge und Raum in der qualitativen Bildungs- und Biographieforschung*. Opladen.
Panowsky, E. (1983). *Meaning in the visual arts*. Chicago.
Pilarzcyk, U., & Mietzner, U. (2005). *Das reflektierte Bild. Die seriell-ikonografische Fotoanalyse in den Erziehungs- und Sozialwissenschaften*. Bad Heilbrunn.
Piper Jaffray Survey. (2014). *Taking Stock with Teens—Fall 2014*. Press release. http://www.piperjaffray.com/2col.aspx?id=287&releaseid=1975435.
Porombka, S. (2016). Sekundentricks. In T. Meyer, J. Dick, P. Moormann, & J. Ziegenbein (Eds.), *Where the magic happens. Bildung nach der Entgrenzung der Kuenste* (pp. 27–35). Munich.
Reckwitz, A. (2015). Die Transformation der Sichtbarkeitsordnungen. *Soziopolis*. https://www.soziopolis.de/beobachten/kultur/artikel/die-transformation-der-sichtbarkeitsordnungen/
Reichert, R. (2008). *Amateure im Netz. Selbstmanagement und Wissentechniken im Web 2.0*. Bielefeld.
Reuter, M., & Koever, C. (2019). Gute Laune und Zensur. https://netzpolitik.org/2019/gute-laune-und-zensur/
Richter, C., & Allert, H. (2017). Poetische Spielzuege als Bildungsoption in einer Kultur der Digitalitaet. In H. Allert, M. Asmussen, & C. Richter (Eds.), *Digitalitaet und Selbst* (pp. 237–262). Bielefeld.
Röll, F.-J. (2014). Social Network Sites. In K. U. Hugger (Ed.), *Digitale Jugendkulturen* (pp. 265–282). Wiesbaden.

Schneider, W. (2015). Dispositive…—überall (und nirgendwo)? Anmerkungen zur Theorei und methodischen Praxis der Dispositivforschung. In J. Othmer & A. Weich (Eds.), *Medien—Bildung—Dispositive* (pp. 21–40). Wiesbaden.

Schreiber, M. (2017). *Digitale Bildpraktiken. Handlungsdimensionen visueller vernetzter Kommunikation*. Unpublished Dissertation, University of Vienna.

Schreiber, M., & Kramer, M. (2016). Verdammt schoen. *Zeitschrift für Qualitative Forschung, 17*(1–2), 81–106.

Turkle, S. (1995). *Life on screen: Identity in the age of Internet*. New York.

Twenge, J. M. (2017). *iGen. Why today's super-connected kids are growing up less rebellios, more tolerant, less happy—And completely unprepared for adulthood*. New York.

van Dijck, J., & Poell, T. (2013). Understanding social media logic. *Media and Communication, 1*(1), 2–14.

Wiedemann, C. (2011). Facebook. Das Assessment-Center der alltaeglichen Lebensfuehrung. In O. Leistert & T. Roehle (Eds.), *Generation Facebook* (pp. 161–181). Bielefeld.

Wueschner, P. (2019). The Internet is dead—Long live the Internet. In R. Muehlhoff, A. Breljak, & J. Slaby (Eds.), *Affekt-Macht-Netz* (pp. 247–290). Bielefeld.

CHAPTER 6

Digital Materiality and Subjectivation: Methodological Aspects of Hybrid Entanglements in Processes of *Bildung*

Patrick Bettinger

Abstract This chapter explores possible connections among discourse analysis, materiality and biographical research in the context of subjectivation. The extant methodological/epistemological concepts linking the Foucauldian idea of discourse with biographical research do not provide clear openings for the incorporation of materiality, specifically those in digital form. This chapter proposes an adapted, modified approach to the analysis of material-discursive practices to the end of investigating the materiality and mediality of relationally understood processes of Bildung. In so doing, it identifies a need for a post-anthropocentric understanding of the biographical that focuses on the variety of socio-medial relational reconfigurations.

Keywords Materiality • Mediality • Biography • Discourse • Relationality

P. Bettinger (✉)
Center for Education and Digital Transformation, Zurich University of Teacher Education, Zurich, Switzerland
e-mail: patrick.bettinger@phzh.ch

P. Bettinger (ed.), *Educational Perspectives on Mediality and Subjectivation*, Palgrave Studies in Educational Media,
https://doi.org/10.1007/978-3-030-84343-4_6

1 Introduction

For many years now, the social sciences, and correspondingly education research, have been interested in the matter of how people become the individuals they are and in the role of society in this process. Accompanying these debates is a continuous dispute around the possible conceptualisation and indeed the value per se of the categories of the 'individual' and 'society'. The mutual interrelatedness of the individual and society has become an accepted commonplace in educational science, albeit one to which the rise of digitalisation appears to pose new challenges due to its transformative multi-level impact. The concept of 'deep mediatisation' (Hepp, 2016) argues that we are currently experiencing a fundamental shift in socio-cultural conditions and media communications whose effects extend into almost all areas of life. The present chapter takes up this discussion in reference to qualitative educational media research as a specific field of education science, illuminating possible interconnections between biographical research and discourse analysis with particular regard to the function of digital artefacts within processes of *Bildung* (see Sect. 2 for further explanation). The methodological considerations underlying the chapter outline possible convergences and lines of connection between biographical research and discourse analysis. In combining these two methodological strands, the contribution argues, we can investigate processes of *Bildung* as specific forms of subjectivation incorporating, in particular, questions of the significance of the material sphere to these processes. While discourse research shows clear references to work analysing subjectivation, qualitative educational research centres the methodology of biographical studies. Accordingly, my concern here will be the extent to which we can relate subjectivation and *Bildung* to each other in theoretical terms, but also what a corresponding interconnection between biographical analysis and discourse analysis might look like.

At the outset of this endeavour, we note a current tendency to conceptualise the analytical strands of biographical research and discourse analysis separately from each other; although both fields of research have seen ongoing development in recent decades, they are as yet effectively without a systematic connection, and their relationship consists largely in mutual incuriosity: "Discourse theory has paid little attention to the relationship between subject and discourse and in particular to the question of what scope for action the subjects, conceptualized as dominated by discourse, may have, while biography research, even at the theoretical level, has not systematically taken discourses into account" (Pohn-Lauggas, 2017, p. 1095). Biographical research has its roots in symbolic interactionism and phenomenology, adhering in the main to an essentialist idea of the subject as capable of rational self-transcendence, while post-structuralist discourse studies, particularly in the succession of Foucault and Butler, hold a non-centralist concept of subjectivation. Approaches drawing on discourse analysis emphasise contextual facets of subjectivation, such as power structures, as the conditions necessarily precedent to the emergence of certain forms of the subject; the focus of biographical research is directed elsewhere, primarily, that is, on the subject's internal absorption, filtration of and reflection on societal structures (Spies & Tuider, 2017a, p. 5). Despite the general consensus on the understanding of biography (in the sense of a life story or life course) as a social construct (Völter et al., 2005, p. 7), biographical research as it takes place empirically is largely, often exclusively, tied to the life stories of individuals, without independent analysis of the social context. Discourse analysis diverges from this in primarily drawing on data collected in a non-reactive manner and, especially in the practice of the associated research, declining, in most instances, to focus on individuals in this form. These differences in research practice and fundamental theoretical orientation (which I outline only roughly here) notwithstanding, recent years have seen the establishment of a number of successful and productive attempts to bring the strands of biographical research and discourse studies together (Pfahl & Traue, 2013; Rose, 2012; Spies, 2009; Spies & Tuider, 2017b; Truschkat, 2018).

One blind spot largely shared by both research directions is the matter of materiality and its significance to sociality. This chapter will tackle the lacuna, arguing that it is precisely in a synthesis of biographical and discourse research that promising openings for materiality appear.

2 *Bildung*, Biography and Discourse: Points of Interconnection and of Departure for Exploring Entanglements of Mediality and Materiality

This section seeks to shed light on the interconnections between this chapter's three central theoretical points of reference: *Bildung*, subjectivation and biographical research. In outlining these moments of linkage, I will generate a theoretical framework as a point of departure for a study of the potential role of materiality, specifically digital artefacts, in processes of *Bildung* and for possible approaches to analysing these processes.

Bildung *and Subjectivation as Complementary Approaches*

The concept of *Bildung*, key to German-language education science, centres in its classical sense on the idea of unfolding potential for human development via an individual's reflection on the self and the world. Originating from the humanist ideal of the subject's self-empowerment initially proposed as a guiding principle for a German education system by Wilhelm von Humboldt in the nineteenth century, *Bildung* has come to signify a way to develop the self by extending the bounds of that self's existing worldview (Koller, 2012, p. 11). Beginning in the 1980s, educationalists in Germany have revisited and advanced the concept; particularly notable developments towards the notion of *Bildung* as a transformative process are in evidence in the work of Rainer Kokemohr (1985) and subsequently of Winfried Marotzki (1990), Hans-Christoph Koller (1999), Heide von Felden (2003) and Arnd-Michael Nohl (2006). In Koller's words, the conceptualisation thus generated "creates a notion of *Bildung* that allows it to be critically linked to both social theory and empirical research" (Koller, 2017, p. 34).

Most of these approaches to *Bildung* are at pains to distinguish it from learning. With reference to Marotzki, Koller (ibid.) asserts that "*learning* should be considered as taking in new information. *Bildung*, on the other hand, would include learning-processes on a higher level, during which the way of *processing* new information changes [...] *Bildung*, then, cannot be understood simply as the process of acquiring knowledge or competencies, but rather as a transformation of the subject's relation to the world, to others and to itself.". This idea of *Bildung* as transformative implies close links to biographical research. There is a consensus within "*Bildungs*-oriented biographical research" (Felden, 2017, p. 153) that the analysis of processes of *Bildung* permits the reconstruction of biographical narrations. Accordingly, the usual procedure is to subject social aspects of *Bildung*, and the influential sets of conditions within which these processes take place, to empirical consideration on the basis of the reconstruction of individual cases, i.e. by analysing narrative interviews (ibid., p. 155). Critics of this approach maintain that it claims to derive supra-individual aspects of biographical processes of transformation near-exclusively from individual narratives (Rosenberg, 2010).

As we continue our survey of the concept, we cannot fail to notice the close association of the concept of *Bildung* with theories of subjectivation. Both of these distinct theoretical strands direct our attention to the interrelationship between the individual and society, figuring it as constitutive of the emergence of the human self-concept and of change therein. Unlike *Bildung*, which frequently posits a subject with the capacity for reflection on itself and the world, the concept of subjectivation, in line with its specific theoretical elaboration in each instance, makes more emphatic reference to external processes via which the outworkings of power fall upon that subject. Notwithstanding the concession that those involved in these processes may well 'have minds of their own' and resist the influence of power as active subjects, theories of subjectivation foreground issues of the subject's constitutedness (as opposed to its (self-)constitution) and the ways in which this subjectivation, as a process both experienced and undergone by individuals, takes place (Ricken, 2019, p. 97). Theories of *Bildung* place contrastingly greater emphasis on forms of emancipatory revolt originating with the individual, which draws their primary attention to the issue of how subjects assert themselves in the face of hegemonic orders, carve out spaces for themselves to exercise their freedom as subjects, and effect change.

Subjectivation is an analytical construct that in many instances refrains from setting normative prescriptions and, although its tendency is to critique power, will frequently approach specific forms of subjectness and manners of subjectivation connotated as 'desirable' with a degree of neutrality. Considerations drawing on theories of *Bildung* differ from this in that, as a rule, they feature, at least implicitly, underlying normative stipulations. Although they do not always reflect on these prescripts or call them into question (Koller, 2016), we may assume the presence of such value-laden criteria in the consensus of thought undergirded by theories of *Bildung*, and perceive their activity in, for example, the demarcation of processes of *Bildung* from other processes and the conferment of pedagogical desirability on the former. *Bildung*, as set out by Taylor (2017, p. 422), is not simply a matter of the acquisition of knowledge or skills, but has a comprehensive regard to profound processes of change against the backdrop of specific ideals: "[Bildung] has been figured as both an intellectual and moral endeavour; it is about more than knowledge, and it is about sensibility and character; and while its focus is the holistic development of the individual, it is also about how individual cultivation is articulated to a vision of a better society. The central concern of Bildung is what constitutes an educated or cultivated human being?".

A further distinction between *Bildung* and subjectivation reveals itself to us as we turn to matters of research and how it is done. Research into subjectivation has recourse to a wide variety of reconstructive and interpretive instruments, with an influential preponderance of poststructuralist conceptualisations; research on *Bildung*, by contrast, specifically that of the qualitative kind and particularly where it emphasises transformational aspects, appears to cluster notably in the area of biographical research (Felden, 2016; Fuchs, 2011; Marotzki, 2006). The predominant methodological technique in evidence is that of narrative interviews whose subjects set out their life stories spontaneously, recounting episodes as they occur to them. The narratives thus attained serve, in biographical research influenced by *Bildung* theories, as a basis for the reconstruction, via a range of methodologies, of potential processes of change in subjects' attitudes to their selves and to the world. The central focus here, then, is on mental processes catalysed by specific life situations: "the transformation of attitude towards the world and the self is mostly initiated by a significant experience and is often connected with reflexivity" (Felden, 2017, p. 158). In this concept, *Bildung* takes place when we observe a transformation in the framing the subject gives to its relationship with itself and the world. The increased flexibility and expansion of individual horizons that *Bildung* entails frequently goes hand in hand with a life episode or life course characterised by the encounter with a crisis for which the attitudes to the self and the world held hitherto fail to offer adequate solutions (ibid.).

This brief outline of the differences between *Bildung* and subjectivation does not tip the scales against the multiple common factors shared by the two, as evident in the recent increase in work which places theoretical considerations around *Bildung* and the subject in reciprocal reference and interrelation (Bünger, 2015; Ricken, 2019; A. Schäfer, 2019) in the hope—seen from the *Bildung* perspective—of, among other things, creating greater space for the analysis of power in theoretical discussions around *Bildung* and raising awareness of the structures within which *Bildung* takes place. This chapter will argue for an interconnection between recent work in subjectivation theory and research (Alkemeyer et al., 2013; Alkemeyer et al., 2018; Geimer et al., 2019) and theories of *Bildung* as a productive way of meeting the concerns of those who perceive a mentalistic reduction of the *Bildung* concept as set out above and, in so doing, call into question the conventional conception of the subject at the centre of theories of *Bildung*, with its Enlightenment tradition drawing on the notion of a subject possessed of the capacity to exercise reason and reflection.

Media pedagogy in particular points to a further convincing argument for linking *Bildung* to subjectivation. In recent years, the latter field has begun to engage with issues related to mediality and materiality, opening up new analytical routes to a nuanced view of how a subject comes to be (Eickelmann, 2019; Geimer & Burghardt, 2019; Spengler, 2018). Analysis of subjectivation that seeks the moment of 'doing', of performativity, provides points of connection to current developments in media theories which conceive of mediality as the "constitutionality of media which reveals that and how they operate" (Jäger, 2015, p. 110) and in so doing set their underlying conception apart from an essentialist view of media, laying greater emphasis on the role of processual human-media configurations (Voss, 2010). The power structures at work in these processes are particularly amenable to explorations via the analytical and terminological toolkit provided to us by research into subjectivation. Media pedagogy which continues to centre the conventional understanding of *Bildung* runs the risk of launching itself from an idea of 'the media' which is essentially a short circuit, adhering to traditional dichotomies and their corresponding prescriptive norms (such as subject/object, body/mind) to an extent that fails to grasp the constitutive power of mediality, particularly in its digital form. The worst case ensuing from this is a view of media as nothing but an accessory to social phenomena which, albeit ubiquitous in their spread and firmly established in our lifeworlds, do not appear as foundationally entangled with processes and paths of subjectivation.

Discursive Traces in Biographical Articulations

Criticism of qualitative empirical research into *Bildung* has held that it frequently—at least in its practical research—focuses too exclusively on the individual, the status of whose narrative, self-reflective statements as the be-all and end-all of processes of *Bildung* in fact merits questioning reassessment. I intend to respond to this admonition and make this aspect of my argument accessible by highlighting the links between biographical and discourse-centred research. In so doing, I proceed from the central assumption that bringing these approaches into interrelationship with one another is productive and, what is more, the synthesis serves to neutralise the weaknesses of each. I will structure this endeavour with a line of argumentation drawing fundamentally on the reciprocal referencing between the two research approaches as outlined above, with a specific spotlight on the relevant methodological issues. It is important to note here that, as far as biographical research in education science is concerned, this bridging of the apparent gap to subjectivation is as yet a preliminary endeavour (Dausien & Hanses, 2016, p. 166).

If we assume the perspective of discourse-centred research, with Foucault as its predominant figure, we may, when examining the basic theoretical premises from which biographical research proceeds, initially find ourselves struggling to perceive where, or indeed that, it offers points of connection. Biographical approaches stem consciously from traditions such as classical interactionism and phenomenology, bearing the key influence of the Chicago School and advancing a corresponding view of the subject (Merrill & West, 2009, p. 22; Völter et al., 2005, p. 10). The concept of the subject put forward by discourse analysis, following Foucault, appears almost diametrically opposed to the biographical view, placing at its heart not the subject in action, but a critical distancing from the conventional notion of the subject with frequent recourse to Foucault's oft-cited proclamation of the subject's 'death'. Schäfer and Völter (2009) have demonstrated that Foucault's critique of the subject does not of necessity imply a rejection of biographical research; instead, they perceive a prompt to the researcher to "*specifically* [*konkret*; emphasis in original] relate the Foucauldian propositions on the origins of the modern subject to the practice of reconstructive biographical research" (ibid., p. 165). Their suggestions for putting this into action include bringing a greater emphasis on performative self-presentation into the analysis of biographical interviews, enabling interrogation of the forms of subject production that become evident to us amid the matrices thus generated. A concomitant of this proposed procedure would be a shift in analytical perspective whose aim would be to recover and expose to a multi-faceted view the relationships to the self that are embedded in articulations around the individual's life course and to identify and note the positions assumed by that self in societal structures of knowledge (ibid., p. 168). This attitude of the researcher would hinge—to put it concisely—on the assumption that discourses permeate the articulation of memory (ibid., p. 171). At heart, we have here a problem for biographical research in the form of a dual discursivity of both everyday experience[1] and the biographical narration that might emerge, for instance, during an interview. There is nothing fundamentally new about this distinction, which appears, for instance, in Rosenthal's (2004) approach to the differentiation of experienced from narrated life stories. If, however, we throw discourse analysis into the mix, and with it the associated questions around societal orders of knowledge, the production and subversion of subject positions or the ways, shapes or forms of subjectivation, we find ourselves facing the additional difficulty of needing to expand our corpus of material with the addition of further types of data; it will, at any rate, appear problematically limiting to train our focus exclusively on narrative interviews if we are to generate assertions which a discourse-analytical perspective would deem robust.

A further observation of relevance here is that of the analysis of power made possible by the forging of an interconnection between research on discourse and biographical work. Discourse analysis opens up a broad methodological path down which we can pursue the assumption of the nature of the life course or 'life story' as a social construct and in so doing undergo a heightening of our awareness of the various forces and inequities within and from whose entanglement subjectivity emerges into being. This path is also a route of access to biographical processes as instances of the re/production, affirmation, transformation or subversion of subjectivations; it broadens our horizons from the subject as an "entity-individual [Individuen-Entität]" (Reckwitz, 2012, p. 16) to the sets of sociocultural conditions framing it. Spies (2009, p. 71) asserts the decisive importance of agency as a concept in the synthesis of discourse-centred and biographical research:

> "because we see time and again in biographical studies that the narrators of life stories, albeit they are interpellated or addressed by discourses, that what they 'make' of the subject positions made available to them—that is, how they fill them, modify them, refuse them—is a great deal more 'wilful', 'autonomous', or just simply more complex and chaotic than the interpellation or address has in mind". (Ibid.)

The dual analytical perspective proposed by Spies in this context sets the frameworks presented to subjects by discourses as distinct from the positions taken by subjects as actors. Following Stuart Hall, Spies proceeds from the premise that interpellation serves as a connector between individuals and discourses in a process of articulation shaped by power which establishes and temporarily stabilises a system constituted of differences (ibid., p. 74). It is via these processes that discourses take on the task of attaching significance and generating a heterogeneous multiplicity of subject positions as effects of this articulation, positions which subjects may assume or leave and which as such create a time-limited identity (ibid., p. 75). We must read this process neither as one of entirely free choice exercised by autonomous subjects nor as a deterministic swallowing up of subjects by all-powerful discourses before they can exercise any agency in the first place. Instead, individuals are called to forge connections within discursive structures they do not have the choice to exit, albeit their constitutive discourses are always incomplete and discursive reconfigurations give rise to a subject's ability to act, that is, to agency—which we can define as the capacity of subjects to take up new subject positions and to invest their resources in them, and therefore as the capacity for self-articulation (ibid., p. 79). Applying these ideas to biographical research, Spies considers that we can, "with the aid of biographical analysis[,] reconstruct positionings [...] that point to specific discourses and document the individual's embeddedness in societal power relations" (ibid., p. 83). Spies shares the perspective identified above on the problem of dual discursivity, observing that discursive positions and positionings are at work both in the moment of narration and in the experienced past. The chief analytical point of reference identified here by Spies as needing careful consideration on the part of the researcher consists in the interference of the discursive embeddedness of the narration in the present, in which it is being produced (ibid., p. 82). This would make the distinguishing characteristic of biographical research inspired by discourse analysis its high capacity to undertake extremely detailed examination of the form taken by subject positionings and the ways in which subjects adopt them.

The considerations set out by Spies doubtless create a significant point of departure for our own train of thought, but we would do an incomplete job at this stage if we were to fail to point out once again the issues with restricting the data analysed to narrative interviews. While a perspective led by biographical analysis may deem such a limitation advantageous, a discourse-analytical view would need to raise the associated significant constraints which, in my view, the inclusion of other types of data material might make more manageable. From this stems the assumption, which will guide the further course of this chapter, that both biographical and discursive processes may manifest in a range of different shapes and forms, including and beyond verbal utterances. What now follows, true to this assumption, will centre the material manifestation of discourses and life stories.

Biographical Processes; a Material-Discursive View

I will follow our exploration of the relationship between *Bildung* and subjectivation, the analytical potential inherent to this relationship, and its methodological translation into a link between biographical analysis and research into discourse with a search for the extent to which this proposal might contain the seeds of a process of analysis driven by awareness of materiality which could fruitfully supplement conventional approaches to biographical analysis. A look at the current state of play confirms that biographical research has paid only marginal attention to materiality to date, and more recent accounts of the concept have no foothold at all in the discipline. Liebsch (2018, p. 45) is among those who point to the lacuna, noting that "[...] we are [...] thus far without attempts to interconnect the 'biographies of things' with [those] of human actors" (Liebsch, 2018, p. 45). The day-to-day practice of biographical research certainly encompasses artefacts, typically documents such as letters, diaries and other autobiographical formats, yet fails by and large to reflect methodologically on the significance of things to biographical processes. As a rule, the recapitulation of life stories within a biographical interview revolves around the spontaneous verbal account. Materiality, or, put differently, socio-material contexts, attract attention, if at all, as components of the narrative, but are not deemed to merit a distinct empirical approach via—for instance—artefact analysis (Lueger & Froschauer, 2018). The conclusion from this observation must be that the material turn (Dolphijn & van der Tuin, 2012; Kissmann & van Loon, 2019), with its emphasis on the significance of material affordances to social contexts, has thus far bypassed biographical research and particularly its methodology, with isolated exceptions (z. B. Engel, 2020; Wundrak, 2015). This gap in the research when it comes to the materiality (and indeed the mediality) of the biographical is not a new discovery (Fetz, 2009), yet few have paid it any note in recent years, with ideas issuing from the area of Actor-Network-Theory (ANT) largely failing to strike a chord in biographical research.

The picture is not greatly different with regard to discourse research, in which engagement with materiality and its significance is yet in its infancy. Notwithstanding a handful of attempts to synthesise, for instance, ANT with dispositive analysis (van Dyk, 2010, 2013), there has been little empirical activity aimed at establishing the significance of material entities in discursive structures. One reason for this may lie in the argument, frequently advanced by the proponents of new materialism, of a rejection of the linguistic turn, which represents a key point of reference at least for many of those discourse-analytical approaches whose chief focus is linguistic. Those schools of discourse analysis centring the sociology of knowledge indeed appear to be drawing very clear lines of demarcation against new materialist approaches, despite supporting in principle an increased emphasis on materiality in discourse analysis (Keller, 2019). New materialist approaches, with their self-set task of taking up and carrying forward the critique of the subject and other concepts first advanced by poststructuralism, would seem to lend themselves to a continuation of discourse analysis as set out in the poststructuralist frame. If we proceed from the conceptualisation of discourse as outlined above with reference to Spies (2009), with its foregrounding of articulation as a discursive practice, we find ourselves able to access considerations around material-discursive practices which also appear, for instance, at the heart of work by Karen Barad (2003). In the words of Schmidt (2019, p. 137), "[m]atter is discursive in the same way that discourse practices are always already material. [...] discourse practices are not activities propped up by humans, but specific material (re-)configurations of the world, each of which enacts boundaries, properties and significance in a different way". Barad's contribution in this context includes the concept of 'intraaction', which she places at the core of her fundamental ontological presumption of the world as a continuous process of becoming, a performative loop of re/configuration. This definition, consciously set at a noticeable distance from the anthropomorphic concept of interaction, serves as Barad's emphasis of her assumption that phenomena constitute themselves by means of their agential potential for discreteness and that relata are therefore not antecedent to the acts of relationising which affirm them as such (Barad, 2003, p. 814):

> "It is through specific agential intra-actions that the boundaries and properties of the "components" of phenomena become determinate and that particular embodied concepts become meaningful. A specific intra-action (involving a specific material configuration of the 'apparatus of observation') enacts an *agential cut* (in contrast to the Cartesian cut—an inherent distinction—between subject and object) effecting a separation between 'subject' and 'object'". (Ibid., p. 815)

Barad, drawing closely on Foucault, conceives of a 'discourse' not as what is said, but as what this 'what-is-said' enables, circumscribes, constrains. In this reading, material-discursive practices are not equivalent to verbal utterances producing meaning, but instead "specific material (re) configurings of the world through which local determinations of boundaries, properties, and meanings are differentially enacted. That is, discursive practices are ongoing agential intra-actions of the world through which local determinacy is enacted within the phenomena produced" (ibid., p. 820).

This view—which we might usefully term post-anthropocentric—of materiality's significance in discursive structures prepares the ground for an analytical heuristic for the interrogation of processes of subjectivation that unfold in the long as well as the shorter term. Practices of a material-discursive nature require retrospective access rather than analysis *in actu*. It is at this point that artefacts, through their status as material manifestations of discursive structures, reveal their central role in the reconstruction of change—understood in the spirit of *Bildung*—in material-discursive configurations.

3 Materiality, Mediality and Discursivity in Processes of 'transformative *Bildung*': Methodological Explorations and Exemplifications

The view of the life course set out above, using the lens of subjectivation theory, adopts the notion of entangled materiality and discursivity as the fundamental process-ontological attitude underlying a relational concept of hybrid forms of being and centres it in its progress through its considerations. What follows will demonstrate the appropriateness and efficacy of this position as a basis for engagement with the mediality undergirding processes of *Bildung*. To do so, we first need to define this mediality, which in my view goes considerably beyond supplying a more rhetorically nuanced term to replace (for example) 'the media'. I see this terminological shift as tied to a corresponding change in media-theory mindset as identified, *inter alia*, by Jäger (2015), who sets out the distinction as follows: 'Media' frequently connotates notions of reification, presuppositions of an availability of these 'media' to external determination of what they *are*. Anchoring one's approach in mediality, by contrast, means asking what media *do*; how they exert impact in practical realities; how they make a difference and take their part in constitutive processes. Mediality, in other words, places its focus on the "operative property of media" (ibid., p. 110), a perception with a performative conceptualisation of media-ness at its root.

Even such a brief definition as this casts a spotlight on points of intersection with the ideas outlined previously in the chapter. A view of biographical processes—of *Bildung*—which proceeds from the notion of configurations and reconfigurations of heterogeneous entities and seeks to identify the characteristic forms of such re/configurations in the material-discursive mode—whose transformation is well described by the concept of *Bildung* itself—will find in the idea of mediality a heuristic expansion of these horizons.[2] This is a perspective which will read digital artefacts not as substantialistic entities, but, following Barad, as material-discursive practices, thus avoiding an over-hasty agential cut in adherence to a traditional subject/object dichotomy. The intent here is to describe, persistently and by small steps, processes of configuration and reconfiguration, holding *Bildung* in mind in order to identify potential transformative instances and variations of it in the course of the process and of time. One potential difficulty crops up, here as in other variants of biographical research, in the inevitable limits imposed by the retrospective mode, which call for a pragmatic approach, backed up by solid methodological foundations, to processes pertaining to the past. A further important factor is the empirical rarity of the profound process of transformation described above as the epitome of *Bildung*, which, far from being the oft-encountered typical case, is instead rather the exception (Geimer, 2014). With this in mind, we would do well to refer to Bourdieu's concept of praxis in reading material-discursive practices as habituated schemata which, while not immutable, tend towards inertia, persistence and the reproduction of extant structures.[3]

The prospect of doing research into *Bildung* on the basis of the approaches we have sought to bring together here prompts us to align ourselves with a post-humanist conception of methodology, which means imparting less attention to the subject as a human being possessed of the capacity for reason and turning instead to the concept of "embodiment—material, affective, finite", which "proves to be of greater importance (ontologically) than consciousness"(Snaza & Weaver, 2015, p. 5). The strength of the humanist tradition is evident testament to the extent of the challenge such a fundamental shift poses to the conceptions education science currently holds of its subject.

Bringing these considerations to bear on the field of research into *Bildung* does not mean we have to expel the biographical narrative interview from our methodological repertoire; it does, however, fundamentally change its status in the research process and lead us inexorably towards the inclusion in our work of additional data types and formats, beyond verbal utterances produced by human actors. Ethnographic methods would seem an obvious choice in this context, yet the long-term character of processes of *Bildung*, and the uncertainty that remains as to whether this methodological approach would succeed in capturing genuinely transformative instances, likely mean they are virtually impossible to apply in practical research.

The totality of the entities involved in processes of reconfiguration is too manifold to offer us realistic prospects of ever completely mapping the field. This raises issues of the assignation of relevance to these entities. An exclusive focus on instances of relevance articulated by human actors would soon see us overrun by the established patterns of anthropocentrism. We might set up one line of defence here by including artefacts in the analysis and in so doing both directing attention towards the material-discursive agency inherent to non-human entities and generating new ways to handle the problem of retrospectivity. I perceive particular promise in this respect in digital artefacts accessible via the internet, such as websites, blogs, social media posts and, fundamentally, any other manifestation of media articulation (cf. Bettinger, 2017, for a concrete methodical implementation). My specific hope here is in their capacity to incorporate 'material agency' (Bettinger, 2018), and its potential for change, into the analysis. This might free us from an exclusive reliance on verbal utterances when we seek to examine the reconfiguration of relational structures. Further, the use (for instance) of internet archives could enable us to examine processes of transformation in the light of changes in artefacts over time. *Bildung* thus appears amenable to examination as a distributed process, whose the analysis can benefit emphatically from not making sole reference to the retrospective linguistic representation of that process, but widening its analytical reach to include the various material manifestations of *Bildung* in their changing relational engagement over time.

This apparent assault on the primacy of verbal utterances in qualitative research is very much a road less travelled, particularly in biographical research, and will require us to rethink foundational concepts and overhaul our terminological toolkit for the description and analysis of processes of *Bildung*, currently stocked typically with products of deeply rooted traditions from the social sciences and humanities. The upshot of these acts of revisiting will be to demonstrate that processes such as thought and action are not the exclusive preserve of human beings, but instead the results of distributed, reciprocally referential instances of relationing among heterogeneous entities (Springgay, 2015, p. 79). Research conducted from this point of view will require acute awareness of the involvement of non-human entities in these entanglements, even and especially where such involvement is not evident at first glance; it will also need the use of an analytical heuristic for the processes of mediation, translation or delegation which unfold in dynamic socio-medial structures. Where our interest lies in *Bildung* conceived of as a transformative process, we will want to pay particular attention to forms of the stabilisation and destabilisation, production and reproduction, and dissolution of relationalities over time. Harnessing the concept of agency outlined above in line with Spies (2017) will enable such an analysis to avoid falling into the trap of rushing to classify specific qualities as human, instead approaching the entities involved in subjectivations with awareness of their multi-layered, heterogeneous constitutions.

Engaging in research under these premises also calls upon us to know they involve a testing of the self-reflection demanded of all researchers engaged in qualitative study. We are called, as prompted by the discipline of science and technology studies (STS), to subject to critical examination our own entanglements with human and non-human actors, which form the preconditions of the emergence of new knowledge, and interrogate them in relation to issues such as effects of power, hegemonic subjectivations within our field of research, and marginalised subject positions. If we are to do genuine justice to the situatedness of the processes that generate knowledge, we need to go beyond methodological expertise and the skilled application of research methods, and adopt into our researcher-being the continuous assessment and reassessment of the agential cuts that we perform in the course of the research process and that constitute the phenomenon at the centre of our interest. And we need to regard this not as an object distinct from and independent of us, but as a process of becoming inextricably interwoven with our practices as researchers.

4 Conclusion: *Bildung*, Biography and the Processuality of Socio-Media Configurations

I hope in this chapter to have highlighted the distinct contrast between the point of view I have laid out and the conventional positions on which biographical analysis typically draws. Further, my intent has been to embark upon an uncertain path whose every step raises questions. In taking up where poststructuralist analyses of discourse have left off, turning to material-discursive practices and to the associated post-anthropocentric and post-essentialist conception of the subject, we might at first glance seem to have lost virtually all connection to biographical research. It has been the business of this chapter to propose a second glance—indeed a more thorough second look which reveals the potential benefits residing in the attempt to bring together these two directions in spite of all the differences in evidence.

Biographisation, in our context and through the lens provided by this chapter, offers itself to a reading as *a mode of the production of continuity in the performative process of relationing*, resting not solely on the life story of an individual as a "structured past, present and future of a life course" (Weidenhaus, 2008, p. 251) but taking in a much wider terrain encompassing the entities involved and seeking the instances of boundary-drawing and agential cuts that, over time, generate continuity and/or discontinuity, patterns of identification and distinction. Having regard to the crisis often acting as the initiating moment of *Bildung* as a process of transformation, we could formulate this phenomenon in terms of a rupturing shock to established ways of relationing, with the potential to catalyse new dynamics in relational structures. The specific value of the joining of these distinct strands of research, as I have proposed here, seems to me to lie in the long-term nature of such processes; as biographical research tends to overlook the significance of material arrangements, so do subjectivation studies and Agential Realism rather neglect long-term developments. There is a distinct difference here from relational analysis, which investigates phenomena *in actu* and whose written observations and video recordings draw on the ethnomethodological tradition. The specific challenge of the biographical brand of material-discursive reconstruction I have put forward here consists in the necessity of drawing retrospective conclusions around configurations of social, media and material entities that had existed—and held relevance—in the past, but no longer exist now.

The appropriateness of the 'biography' label, with its highly anthropomorphic inherent connotations, to the sort of work imagined here is perhaps arguable. We at least find ourselves unable to escape the insight that our proposition requires a significant expansion of the concept of the nature of 'life stories' as social constructs (Alheit & Dausien, 2000), an expansion which would incorporate non-human entities, as independent and possessed of agency, into every stage of the analysis. Indeed, it seems we will need to consider a life story as materially/discursively constructed. A conception of *Bildung* from the point of view of a "decentring of the subject" (Koller, 2001) would be a necessary, but not sufficient precondition of research approached in this way.

I will conclude by noting the pressing need for a debate on the ethical and normative principles which, from the perspective of education science, we cannot do without. If we as researchers attempt to proceed as I have invited us to do in this chapter—in a manner representing a marked departure from the humanist notions underlying the current conception of *Bildung*—we face numerous questions with fundamental salience, not solely to education research, but to the foundations of the education science we know today and the identity of our discipline—indeed questions with the potential to turn these on their head. I believe we are challenged to take up this stimulus and carry it into future research, if we are to remain productively relevant to the academic discourse in this area and retain our capacity to analyse, critique and influence the referentiality to the practical world of action that is so characteristic of education science as a discipline.

Notes

1. In the interest of avoiding misunderstandings, I would like to emphasise at this point that Foucault and the bulk of the discourse analysis-centred approaches following him do not hold the notion of a one-sided, autonomously active discursive sphere confronting subjects, but rather regard discursive practices as effectively the 'living core' from which discursive phenomena issue. Some theoreticians of discourse analyses, however, may disagree.
2. Cf. also the introduction to this volume.
3. A counter-position emphasising the instability of practice is set out in (H. Schäfer, 2013).

References

Alheit, P., & Dausien, B. (2000). Die biographische Konstruktion der Wirklichkeit. Überlegungen zur Biographizität des Sozialen. In E. M. Hoerning (Ed.), *Biographische sozialisation* (pp. 257–284). Lucius & Lucius.

Alkemeyer, T., Bröckling, U., & Peter, T. (Eds.). (2018). *Jenseits der Person. Zur Subjektivierung von Kollektiven.* transcript.

Alkemeyer, T., Budde, G., & Freist, D. (Eds.). (2013). *Selbst-Bildungen. Soziale und kulturelle Praktiken der Subjektivierung.* transcript.

Barad, K. (2003). Posthumanist performativity: Toward an Understanding of how matter comes to matter. *Signs: Journal of Women in Culture and Society, 28*(3), 801–831.

Bettinger, P. (2017). Rekonstruktive Medienbildungsforschung – Die Analyse von Bildungsprozessen als Habitustransformationen in mediatisierten Lebenswelten. http://publ.forschungswerkstatt-medienpaedagogik.de/

Bettinger, P. (2018). *Praxeologische Medienbildung. Theoretische und empirische Perspektiven auf sozio-mediale Habitustransformationen.* Springer VS.

Bünger, C. (2015). "Die Frage der Mündigkeit". Einsätze der Kritik in Bildungstheorie und Bildungsforschung. *Vierteljahrsschrift für wissenschaftliche Pädagogik, 91*(4), 457–474.

Dausien, B., & Hanses, A. (2016). Konzeptualisierungen des Biografischen – Zur Aktualität biografiewissenschaftlicher Perspektiven in der Pädagogik. *Zeitschrift für Pädagogik, 62*(2), 159–171.

Dolphijn, R., & van der Tuin, I. (2012). *New materialism: Interviews & cartographies.* Open Humanities Press.

Eickelmann, J. (2019). Mediatisierte Missachtung als Modus der Subjektivation. Eine intraaktive, performativitätstheoretische Perspektive auf die Kontingenz internetbasierter Adressierungen. In A. Geimer, S. Amling, & S. a. Bosančić (Eds.), *Subjekt und Subjektivierung. Empirische und theoretische Perspektiven auf Subjektivierungsprozesse* (pp. 169–190). Wiesbaden.

Engel, J. (2020). The materiality of biographical knowledge – Ethnography on aesthetic practices. *Ethnography and Education, 15*(3), 377–393. https://doi.org/10.1080/17457823.2020.1761416

Felden, H. v. (2003). *Bildung und Geschlecht zwischen Moderne und Postmoderne. Zur Verknüpfung von Bildungs-, Biographie-und Genderforschung.* Springer VS.

Felden, H. v. (2016). Zum Verhältnis von Bildungstheorie und bildungsbiographischer empirischer Forschung. In D. Verständig, J. Holze, & R. Biermann (Eds.), *Von der Bildung zur Medienbildung* (pp. 89–106). Springer.

Felden, H. V. (2017). Transformative learning, *Bildung* and biographical research. In A. Laros, T. Fuhr, & E. W. Taylor (Eds.), *Transformative Learning Meets Bildung. An international exchange* (pp. 153–162). Sense Publishers

Fetz, B. (2009). Der Stoff, aus dem das (Nach-) Leben ist. Zum Status biographischer Quellen. In B. Fetz (Ed.), *Die Biographie – Zur Grundlegung ihrer Theorie* (pp. 103–154). de Gruyter.

Fuchs, T. (2011). *Bildung und Biographie. Eine Reformulierung der bildungstheoretisch orientierten Biographieforschung*. transcript.

Geimer, A. (2014). Zur Unwahrscheinlichkeit von Bildung. Potenzielle Subjektivierungskrisen vor dem Hintergrund der Relation von Habitus, Identität und diskursiven Subjektfiguren. In F. v. Rosenberg & A. Geimer (Eds.), *Bildung unter Bedingungen kultureller Pluralität* (pp. 195-213). : Springer VS.

Geimer, A., Amling, S., & Bosančić, S. a. (Eds.). (2019). *Subjekt und Subjektivierung. Empirische und theoretische Perspektiven auf Subjektivierungsprozesse*. Springer VS.

Geimer, A., & Burghardt, D. (2019). Die Mediatisierung von Subjektivierungsprozessen. Geschlechternormen im Kontext der Subjektnorm des disziplinierten Selbst in YouTube-Videos und mimetische Praktiken der Subjektivierung. In A. Geimer, S. Amling, & S. a. Bosančić (Eds.), *Subjekt und Subjektivierung. Empirische und theoretische Perspektiven auf Subjektivierungsprozesse* (pp. 235–258). Springer VS.

Hepp, A. (2016). Pioneer communities: Collective actors in deep mediatisation. *Media, Culture & Society, 38*(6), 918–933. https://doi.org/10.1177/0163443716664484

Jäger, L. (2015). Medialität. In E. Felder & A. Gardt (Eds.), *Handbuch Sprache und Wissen*. de Gruyter.

Keller, R. (2019). New materialism? A view from sociology of knowledge. In U. T. Kissmann & J. van Loon (Eds.), *Discussing new materialism. Methodological implications for the study of materialities* (pp. 151–172). Springer VS.

Kissmann, U. T., & van Loon, J. (Eds.). (2019). *Discussing new materialism. Methodological implications for the study of materialities*. Springer VS.

Kokemohr, R. (1985). Modalisierung und Validierung in schulischen Lehr- Lern-Prozessen. In R. Kokemohr & W. Marotzki (Eds.), *Interaktionsanalysen in pädagogischer Absicht* (pp. 177–235). Peter Lang.

Koller, H.-C. (1999). *Bildung und Widerstreit: Zur Struktur biographischer Bildungsprozesse in der (Post-) Moderne*. Wilhelm Fink.

Koller, H.-C. (2001). Bildung und die Dezentrierung des Subjekts. In B. Fritzsche, A. Schmidt, J. Hartmann, & A. Tervooren (Eds.), *Dekonstruktive Pädagogik. Erziehungswissenschaftliche Debatten unter poststrukturalistischen Perspektiven* (pp. 35–48). Leske + Budrich.

Koller, H.-C. (2012). *Bildung anders denken. Einführung in die Theorie transformatorischer Bildungsprozesse*. Kohlhammer.

Koller, H.-C. (2016). Ist jede Transformation als Bildungsprozess zu begreifen? In D. Verständig, J. Holze, & R. Biermann (Eds.), *Von der Bildung zur Medienbildung* (pp. 149–161). Springer VS.

Koller, H.-C. (2017). *Bildung* as a transformative process. In A. Laros, T. Fuhr, & E. W. Taylor (Eds.), *Transformative Learning Meets Bildung. An international exchange* (pp. 33–42). Sense Publishers.

Liebsch, K. (2018). Konstruktivismus und Biographieforschung. In H. Lutz, M. Schiebel, & E. Tuider (Eds.), *Handbuch Biographieforschung* (pp. 37–47). Springer VS.

Lueger, M., & Froschauer, U. (2018). *Artefaktanalyse. Grundlagen und Verfahren*. Springer VS.

Marotzki, W. (1990). *Entwurf einer strukturalen Bildungstheorie. Biographietheoretische Auslegung von Bildungsprozessen in hochkomplexen Gesellschaften*. Deutscher Studienverlag.

Marotzki, W. (2006). Bildungstheorie und allgemeine Biographieforschung. In H.-H. Krüger & W. Marotzki (Eds.), *Handbuch erziehungswissenschaftliche Biographieforschung* (2nd. ed., pp. 59–70). VS.

Merrill, B., & West, L. (2009). *Using biographical methods in social research*. Sage.

Nohl, A.-M. (2006). *Bildung und Spontaneität. Phasen biographischer Wandlungsprozesse in drei Lebensaltern – Empirische Rekonstruktionen und pragmatistische Reflexionen*. Budrich.

Pfahl, L., & Traue, B. (2013). Die Erfahrung des Diskurses. Zur Methode der Subjektivierungsanalyse in der Untersuchung von Bildungsprozessen. In R. Keller & I. Truschkat (Eds.), *Methodologie und Praxis der Wissenssoziologischen Diskursanalyse. Band 1: Interdisziplinäre Perspektiven* (Vol. 1, pp. 425–450). Springer VS.

Pohn-Lauggas, M. (2017). Biography and discourse: A biography and discourse analysis combining case study on women's involvement in National Socialism. *Current Sociology, 65*(7), 1094–1111.

Reckwitz, A. (2012). *Subjekt* (3. Aufl. ed.). transcript.

Ricken, N. (2019). Bildung und Subjektivierung. Bemerkungen zum Verhältnis zweier Theorieperspektiven. In N. Ricken, R. Casale, & C. Thompson (Eds.), *Subjektivierung. Erziehungswissenschaftliche Theorieperspektiven* (pp. 95–118). Beltz Juventa.

Rose, N. (2012). Subjekt, Bildung, Text. Diskurstheoretische Anregungen und Herausforderungen für biographische Forschung. In I. Miethe & H.-R. Müller (Eds.), *Qualitative Bildungsforschung und Bildungstheorie* (pp. 111–126). Budrich.

Rosenberg, F. v. (2010). Bildung und das Problem der Weltvergessenheit. Überlegungen zu einer empirisch fundierten Bildungstheorie im Anschluss an Pierre Bourdieu. *Vierteljahresschrift für wissenschaftliche Pädagogik, 86*, 571–586.

Rosenthal, G. (2004). Biographical research. In C. Seale, G. Gobo, J. F. Gubrium, & D. Silverman (Eds.), *Qualitative research practice* (pp. 48–64). Sage.

Schäfer, A. (2019). Bildung und /als Subjektivierung. Annäherungen an ein schwieriges Verhältnis. In N. Ricken, R. Casale, & C. Thompson (Eds.),

Subjektivierung. Erziehungswissenschaftliche Theorieperspektiven (pp. 119–138). Beltz Juventa.

Schäfer, H. (2013). *Die Instabilität der Praxis. Reproduktion und Transformation des Sozialen in der Praxistheorie*. Velbrück Wissenschaft.

Schäfer, T., & Völter, B. (2009). Subjekt-Positionen. Michel Foucault und die Biographieforschung. In B. Völter, B. Dausien, H. Lutz, & G. Rosenthal (Eds.), *Biographieforschung im Diskurs* (2nd. ed., pp. 161-188). VS.

Schmidt, R. (2019). Materiality, meaning, social practices: Remarks on new materialism. In U. T. Kissmann & J. van Loon (Eds.), *Discussing new materialism. Methodological implications for the study of materialities* (pp. 135–149). Springer VS.

Snaza, N., & Weaver, J. A. (2015). Introduction. Education and the posthumanist turn. In N. Snaza & J. A. Weaver (Eds.), *Posthumanism and educational research* (pp. 1–14). Routledge.

Spengler, A. (2018). *Das Selbst im Netz. Zum Zusammenhang von Sozialisation, Subjekt, Medien und ihren Technologien*. Ergon.

Spies, T. (2009). Diskurs, Subjekt und Handlungsmacht. Zur Verknüpfung von Diskurs- und Biografieforschung mithilfe des Konzepts der Artikulation [70 Absätze]. *Forum Qualitative Sozialforschung, 10*.

Spies, T. (2017). Subjektpositionen und Positionierungen im Diskurs. Methodologische Überlegungen zu Subjekt, Macht und Agency im Anschluss an Stuart Hall. In T. Spies & E. Tuider (Eds.), *Biographie und Diskurs. Methodisches Vorgehen und Methodologische Verbindungen* (pp. 69–90). Springer VS.

Spies, T., & Tuider, E. (2017a). Biographie und Diskurs – eine Einleitung. In T. Spies & E. Tuider (Eds.), *Biographie und Diskurs. Methodisches Vorgehen und Methodologische Verbindungen* (pp. 1–20). Springer VS.

Spies, T., & Tuider, E. (Eds.). (2017b). *Biographie und Diskurs. Methodisches Vorgehen und Methodologische Verbindungen*. Springer VS.

Springgay, S. (2015). "Approximate-Rigorous Abstractions". Propositions of activation for posthumanist research in education. In N. Snaza & J. A. Weaver (Eds.), *Posthumanism and educational research* (pp. 76–88). Routledge.

Taylor, C. A. (2017). Is a posthumanist Bildung possible? Reclaiming the promise of Bildung for contemporary higher education. *Higher Education, 47*, 419–435.

Truschkat, I. (2018). Diskurstheoretische Ansätze der Biographieforschung. In H. Lutz, M. Schiebel, & E. Tuider (Eds.), *Handbuch Biographieforschung* (pp. 127–138). Springer VS.

van Dyk, S. (2010). Verknüpfte Welt *oder* Foucault meets Latour. Zum Dispositiv als Assoziation. In R. Feustel & M. Schochow (Eds.), *Zwischen Sprachspiel und Methode. Perspektiven der Diskursanalyse* (pp. 169–196). transcript.

van Dyk, S. (2013). Was die Welt zusammenhält. Das Dispositiv als Assoziation und performative Handlungsmacht. *Zeitschrift für Diskursforschung, 1*, 46–66.

Völter, B., Dausien, B., Lutz, H., & Rosenthal, G. (2005). Einleitung. In B. Völter, B. Dausien, H. Lutz, & G. Rosenthal (Eds.), *Biographieforschung im Diskurs* (pp. 7–20).

Voss, C. (2010). Auf dem Weg zu einer Medienphilosophie anthropomedialer Relationen. *Zeitschrift für Medien- und Kulturforschung (ZMK), 2*, 169–184.

Weidenhaus, G. (2008). Modi der Biographisierung. *Sozialer Sinn, 9*(2), 251–279.

Wundrak, R. (2015). Die Materialität des Erzählens. Die Bedeutung von Dingen und Körpern in einem biographischen Interview. Ein Beispiel aus Jaffa (Israel). *Österreichische Zeitschrift für Soziologie, 40*, 355–371.

Index[1]

[1] Note: Page numbers followed by 'n' refer to notes.

P. Bettinger (ed.), *Educational Perspectives on Mediality and Subjectivation*, Palgrave Studies in Educational Media,
https://doi.org/10.1007/978-3-030-84343-4_6

The manufacturer's authorised representative in the EU is Springer Nature Customer Service Centre GmbH, Europaplatz 3, 69115 Heidelberg, Germany. If you have any concerns regarding our products, please contact ProductSafety@springernature.com

Printed and bound by CPI Group (UK) Ltd, Croydon, CR0 4YY
30/09/2026
02225711-0003